Lecture Notes in Mathematics

Edited by A. Dold and B. Eckmann

813

Antonio Campillo

Algebroid Curves in Positive Characteristic

Springer-Verlag
Berlin Heidelberg New York 1980

Author

Antonio Campillo
Departamento de Algebra y Fundamentos, Facultad de Ciencias,
Universidad de Valladolid
Valladolid/Spain

AMS Subject Classifications (1980): 14B05, 14H20

ISBN 3-540-10022-9 Springer-Verlag Berlin Heidelberg New York
ISBN 0-387-10022-9 Springer-Verlag New York Heidelberg Berlin

Printing and binding: Beltz Offsetdruck, Hemsbach/Bergstr.
2141/3140-543210

INTRODUCTION

A number of definitions of equisingularity have appeared, since Zariski published his "Studies in Equisingularity". Those definitions are equivalent in the particular case of plane algebroid curves over an algebraically closed field of characteristic zero (the situation that Zariski considered initially). However the case of characteristic $p > 0$ has not received extensive attention and only a few papers are available: Lejeune [15], Moh [18], and more recently Angermüller [3].

These notes intend to give a systematic development of the theory of equisingularity of irreducible algebroid curves over an algebraically closed field of arbitrary characteristic, using as main tool the Hamburger-Noether expansion instead of the Puiseux expansion which is usually employed in characteristic zero.

The so called Hamburger-Noether expansion first appeared as an attempt to obtain parametrizations of plane algebraic curves in any characteristic. It was completely developed in a work by G. Ancochea, published in Acta Salamanticiensis (Universidad de Salamanca) and not available any longer. Essentially it is based on a parametrization of an irreducible algebroid curve $\square = k[[x,y]]$ over k of the type

$$x = x(z_r)$$
$$y = y(z_r) \quad ,$$

z_r being an element of the quotient field of $\square$, obtained from x, y by a chain of relations

$$y = a_{01}x + a_{02}x^2 + \dots + a_{0h}x^h + x^h z_1$$
$$x = a_{12}z_1^2 + \dots + a_{1h_1}z_1^{h_1} + z_1^{h_1} z_2$$
$$\dots\dots\dots\dots\dots\dots\dots\dots\dots\dots$$
$$z_{r-1} = a_{r2}z_r^2 + \dots\dots\dots\dots$$

where $a_{ji} \in k$.

This expansion enables us to define a system of characteristic exponents of a plane curve which is equivalent to that derived from the Puiseux expansion in characteristic zero. These exponents determine and are determined by the resolution process for the singularity of the curve, by the semigroup of values of its local ring, etc...

Chapter I contains well known definitions and results on algebroid curves, existence of parametrizations, and resolution of singularities of an irreducible algebroid curve. Chapter II is devoted to the Hamburger-Noether expansion and comparison of it with the Puiseux expansion in characteristic zero.

In chapter III, by using a complex model for the singularity of a curve over an algebraically closed field of any characteristic, we introduce the characteristic exponents. From this model we compare these exponents with the usual ones, and compute them in terms of Hamburger-Noether expansions and Newton polygons.

The semigroup of values of the local ring of the curve is calculated from the values of the maximal contact or from the characterisitic exponents in chapter IV. We also find the relationship between the characteristic exponents and the Newton coefficients given by Lejeune.

Finally, in chapter V we study several criteria for equisingularity of irreducible twisted curves.

I would like to express my sincere thanks to Professor Aroca for his comments and suggestions.

TABLE OF CONTENTS

CHAPTER I

PARAMETRIZATIONS OF ALGEBROID CURVES

This chapter is devoted to the systematization of the concept of local parametrization of irreducible algebroid curves over an algebraically closed ground field of any characteristic. Although there are not essential differences with the case of characteristic zero, we have thought it useful to treat this case in detail.

1. PRELIMINARY CONCEPTS.

Let k be an algebraically closed field of arbitrary characteristic. If $\underline{X} = \{X_i\}_{1 \leqslant i \leqslant N}$ is a set of indeterminates over k, we shall denote by $k((\underline{X})) = k((X_1, \ldots, X_N))$ the formal power series ring in the indeterminates $\underline{X}$ with coefficients in k. The function order on $k((\underline{X}))$ will be denoted by $\underline{\upsilon}$.

The Weierstrass Preparation Theorem (W.P.T.) will be used frecuently in this work. It is stated down and its proof and direct consequences may be found in Zariski-Samuel, [29].

Theorem 1.1.1.(W.P.T.).- Let $f(\underline{X}) \in k((\underline{X}))$ be a series which is regular in X_N of order s, i.e., $s = \underline{\upsilon}(f(0,\ldots,0,X_N))$. Then, there exist a unique unit $U(\underline{X})$ in $k((\underline{X}))$ and a unique degree s monic polynomial $P((\underline{X}'),X_N)$ (where $\underline{X}' = (X_1,\ldots,X_{N-1})$) in X_N with coefficients in $k((\underline{X}'))$ such that

$$f(\underline{X}) = U(\underline{X}) \cdot P((\underline{X}'),X_N).$$

<u>Definition 1.1.2.</u>- An <u>irreducible algebroid curve</u> (or simply a <u>curve</u> if there is no confusion) over k is a noetherian local domain $\square$ such that:

1) $\square$ is complete.

2) $\square$ has Krull dimension 1.

3) k is a coefficient field for $\square$.

If $\underline{m}$ is the maximal ideal of $\square$, the property 3) means that k is contained in $\square$ and is isomorphic to the field $\square/_{\underline{m}}$ by the canonical epimorphism $\square \longrightarrow \square/_{\underline{m}}$.

<u>Remark 1.1.3.</u>- Since $\square$ is noetherian, the vector space $\underline{m}/_{\underline{m}}2$ over k is finite dimensional. The number $\mathrm{Emb}(\square) = \dim_k(\underline{m}/_{\underline{m}}2)$ is called the <u>embedding dimension of $\square$</u>.

For every basis $B = \{x_i\}_{1 \leqslant i \leqslant N}$ of the maximal ideal $\underline{m}$,

$$S = \{ x_i + \underline{m}^2 \}_{1 \leqslant i \leqslant N}$$

is a set of generators of the k- vector space $\underline{m}/_{\underline{m}}2$. The set S becomes a basis of this vector space if and only if B is a minimal basis of the ideal $\underline{m}$.

Let $B = \{x_i\}_{1 \leqslant i \leqslant N}$ a basis of $\underline{m}$. Using 3) and the completeness of $\square$ we find a natural surjective k-homomorphism

$$(1) \qquad \begin{array}{ccc} k((\underline{X})) & \longrightarrow & \square \\ X_i & \longmapsto & x_i \end{array} \qquad , \quad \underline{X} = \{X_i\}_{1 \leqslant i \leqslant N} .$$

Thus, there exists an isomorphism $k((\underline{X}))/_{\underline{p}} \cong \square$, where $\underline{p}$ is a prime ideal of $k((\underline{X}))$. The condition 2) means that the depth of $\underline{p}$ is 1.

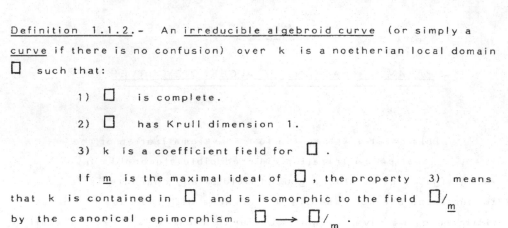

We may identify $\square$ with the ring $k((\underline{X}))/_{\underline{p}}$. In fact, if we set $x_i = X_i + \underline{p}$, we can write $\square = k((x_1, \ldots, x_N))$. The minimun N such that these isomorphisms exist, is exactly $\mathrm{Emb}(\square)$. When an identification as above is done, we shall say that *the curve is embedded in an N-space*. Thus, to give an embedded curve C is to give a prime ideal $\underline{p} \subset k[[\underline{X}]]$ of depth 1. Then, *□ is called the local ring of C*. *The word "embedding" has a precise meaning in scheme theory: The N-space is by definition the affine scheme Spec(k[[X]]), the curve □ is identified with Spec(□) and the homomorphism (1) induces a closed embedding of schemes Spec(□) $\longrightarrow$ Spec(k[[$\underline{X}$]]).*

Now we shall give the following normalization theorem which allows to simplify the form of the ideal $\underline{p}$. Notice the assumptions that $\underline{p}$ is prime and of depth 1, which do not affect the proof. Therefore, we shall prove it for any ideal $\underline{a}$ of a formal power series ring.

Theorem 1.1.4.- Let $\underline{Y} = \{Y_i\}_{1 \leq i \leq N}$, $\underline{X} = \{X_i\}_{1 \leq i \leq N}$ be two sets of indeterminates over k and $\underline{a}'$ an ideal of $k[[\underline{Y}]]$, $\underline{a}' \neq (0),(1)$. There exists an integer m, $0 \leq m \leq N-1$, and an isomorphism Φ from $k[[\underline{Y}]]$ onto $k[[\underline{X}]]$ defined by linear relations $\Phi(Y_i) = L_i(\underline{X})$, $1 \leq i \leq N$, independent over k, such that the ideal $\underline{a} = \Phi(\underline{a}')$ has the following properties:

1) $\underline{a} \cap A_m = (0)$, $\underline{a} \cap A_i \neq (0)$, $m+1 \leq i \leq N$;

   where $A_i = k[[X_1, \ldots, X_i]]$, $0 \leq i \leq N$.

2) There are $N-m$ non zero series $f_i(X_1, \ldots, X_i) \in \underline{a} \cap A_i$, $m+1 \leq i \leq N$, such that

$$\underline{v}(f_i(0, \ldots, 0, X_i)) = \underline{v}(f_i(X_1, \ldots, X_i)) .$$

Proof: We shall construct by induction the isomorphism Φ .

First, let us construct an isomorphism $\Phi^{(1)}$ from $k[[\underline{Y}]]$ onto $k[[\underline{Y}^{(1)}]]$, with $\underline{Y}^{(1)} = \{Y_i^{(1)}\}_{1 \leq i \leq N}$ new variables, such that if $\underline{a}^{(1)} = \Phi^{(1)}(\underline{a}')$, then there exists $f_N^{(1)} \in \underline{a}^{(1)}$ verifying $\underline{v}(f_N^{(1)}(0, \ldots, 0, Y_N^{(1)})) = \underline{v}(f_N^{(1)}(Y_1^{(1)}, \ldots, Y_N^{(1)}))$.

To prove this, take $f'_N \in \underline{a}'$, $f'_N \neq 0$. Let $f_{N,q}$ the leading form of f'_N . If $f_{N,q}(0, \ldots, 0, Y_N) \neq 0$ we set $Y_i^{(1)} = Y_i$, $1 \leq i \leq N$,

$\Phi^{(1)}$ the identity in $k[[\underline{Y}]]$, and $f_N = f'_N$. If $f_{N,q}(0,\ldots,0,Y_N)=0$ we pick out $a_i^{(1)} \in k$, $1 \leq i \leq N-1$, such that $f_{N,q}(a_1^{(1)},\ldots,a_{N-1}^{(1)},1)\neq 0$ (Hilbert's Nullstellensatz). Then, the linear forms

$$L_i^{(1)} = Y_i^{(1)} + a_i^{(1)} Y_N^{(1)} , \quad 1 \leq i \leq N-1$$
$$L_N^{(1)} = Y_N^{(1)}$$

are lineary independent over k and the isomorphism

$$\Phi^{(1)} : k[[\underline{Y}]] \longrightarrow k[[\underline{Y}^{(1)}]]$$
$$Y_i \longmapsto L_i^{(1)}(\underline{Y}^{(1)})$$

and the series $f_N^{(1)} = \Phi^{(1)}(f'_N) \in \underline{a}^{(1)}$ verify our conditions.

Now, let p be an integer, $1 \leq p < N$, and $\underline{Y}^{(p)} = \{Y_i^{(p)}\}_{1 \leq i \leq N}$ indeterminates over k. Assume that there exists an isomorphism $\Phi^{(p)} : k[[\underline{Y}]] \longrightarrow k[[\underline{Y}^{(p)}]]$ given by lineary independent forms such that, if $\underline{a}^{(p)} = \Phi^{(p)}(\underline{a}')$, there exist non zero series $f_i^{(p)} \in \underline{a}^{(p)} \cap k[[Y_1^{(p)},\ldots,Y_i^{(p)}]]$, $N-(p-1) \leq i \leq N$, such that

$$\underline{\upsilon}(f_i^{(p)}(Y_1^{(p)},\ldots,Y_i^{(p)})) = \underline{\upsilon}(f_i^{(p)}(0,\ldots,0,Y_i^{(p)})).$$

Assume that

$$(1) \qquad \underline{a}^{(p)} \cap k[[Y_1^{(p)},\ldots,Y_{N-p}^{(p)}]] \neq (0).$$

Then, take a non zero series in $\underline{a}^{(p)} \cap k[[Y_1^{(p)},\ldots,Y_{N-p}^{(p)}]]$ and use the above procedure to find an appropriate linear change which gives rise to an isomorphism $\Phi^{(p+1)} : k[[\underline{Y}]] \longrightarrow k[[\underline{Y}^{(p+1)}]]$ such that, if $\underline{a}^{(p+1)} = \Phi^{(p+1)}(\underline{a}')$, then there exist non zero series $f_i^{(p+1)} \in \underline{a}^{(p+1)} \cap k[[Y_1^{(p+1)},\ldots,Y_i^{(p+1)}]]$, $N-p \leq i \leq N$, verifying

$$\underline{\upsilon}(f_i^{(p+1)}(Y_1^{(p+1)},\ldots,Y_i^{(p+1)})) = \underline{\upsilon}(f_i^{(p+1)}(0,\ldots,0,Y_i^{(p+1)})).$$

Since $\underline{a}' \neq (1)$ we have $\underline{a}' \cap k = (0)$, hence there exists an integer p such that (1) does not hold. Setting $m=N-p$, $\underline{X} = \underline{Y}^{(p)}$, $\Phi = \Phi^{(p)}$ and $f_i = f_i^{(p)}$ the conditions stated in the theorem are trivially true.

<u>Remark 1.1.5.</u>- In the above theorem if $g_i \in A_i$ denotes an irreducible series which divides f_i, then

$$\underline{\upsilon}\,(g_i(0,\dots,0,X_i)) = \underline{\upsilon}\,(g_i(X_1,\dots,X_i)).$$

Thus if $\underline{a}'$ is prime, the series f_i may be chosen to be irreducible.

Let $\square$ be a complete local ring (for its $\underline{m}$-adic topology, where $\underline{m}$ is its maximal ideal). Suppose that k is a coefficient field of $\square$. For any finite set $\{z_i\}_{1 \leqslant i \leqslant N} \subset \underline{m}$ and indeterminates $\underline{Z} = \{Z_i\}_{1 \leqslant i \leqslant N}$ there is a homomorphism

$$\phi : k((\underline{Z})) \longrightarrow \square$$

given by $\phi(Z_i) = z_i$, $1 \leqslant i \leqslant N$, which is continuous for their respective Krull topologies.

<u>Definition 1.1.6.</u>- We say that $\{z_i\}_{1 \leqslant i \leqslant N}$ are <u>formally independent over k</u> if the above homomorphism is injective.

<u>Theorem 1.1.7.</u>- Let $\underline{X} = \{X_i\}_{1 \leqslant i \leqslant N}$ be indeterminates over k, $\underline{a}$ an ideal of $k((\underline{X}))$ and m an integer, $0 \leqslant m \leqslant N-1$, such that:

(a) $\underline{a} \cap A_m = (0)$, $\underline{a} \cap A_i \neq (0)$, $m+1 \leqslant i \leqslant N$.

(b) There exist non zero series $f_i \in \underline{a} \cap A_i$, $m+1 \leqslant i \leqslant N$, such that

$$\underline{\upsilon}\,(f_i(0,\dots,0,X_i)) = \underline{\upsilon}(f_i(X_1,\dots,X_i)).$$

Set $\square = k((\underline{X}))/_{\underline{a}}$, $x_i = X_i + \underline{a}$, and denote by $\underline{M}$ (resp. $\underline{m}$) the maximal ideal of $k((\underline{X}))$ (resp. $\square$). Then the following statements are true:

1) If $m > 0$,

i) $\{x_i\}_{1 \leq i \leq m}$ are formally independent over k.

ii) $\square = k((x_1, \ldots, x_m))(x_{m+1}, \ldots, x_N)$.

iii) $\square$ is an integral extension of $k((x_1, \ldots, x_m))$.

iv) The height of $\underline{a}$ is $N-m$, and hence its depth is m. Particulary $m = \dim(\square)$.

2) $m = 0$ if and only if $\underline{a}$ is a M-primary ideal.

<u>Proof:</u>

1) Case $m > 0$.

i) The canonical homomorphism $\phi : k((X_1, \ldots, X_m)) \longrightarrow \square$ given by $\phi(X_i) = x_i$ is injective because $\underline{a} \cap A_m = (0)$.

ii) As $k((x_1, \ldots, x_m))$ is a subring of $\square$, we have

$$k((x_1, \ldots, x_m))(x_{m+1}, \ldots, x_N) \subset \square .$$

Conversely, if $f(\underline{X}) \in k((\underline{X}))$, by using the properties of the series f_i we may apply the division algorithm (Zariski-Samuel, (29)), and write

$$f(X_1, \ldots, X_N) = \sum_{i=m+1}^{N} U_i(X_1, \ldots, X_N)\, f_i(X_1, \ldots, X_i) +$$

$$+ \sum_{i_N=0}^{q_N-1} \cdots \sum_{i_{m+1}}^{q_{m+1}-1} R_{i_N, \ldots, i_{m+1}}(X_1, \ldots, X_m) X_{m+1}^{i_{m+1}} \cdots X_N^{i_N} ,$$

where $U_i(X_1, \ldots, X_N) \in k((\underline{X}))$; $R_{i_N, \ldots, i_{m+1}}(X_1, \ldots, X_m) \in k((X_1, \ldots, X_m))$ and $q_i = \underline{v}(f_i(X_1, \ldots, X_i))$. We obtain

$$f(x_1, \ldots, x_N) = \sum_{i_N=0}^{q_N-1} \cdots \sum_{i_{m+1}=0}^{q_{m+1}-1} R_{i_N, \ldots, i_{m+1}}(x_1, \ldots, x_m)\, x_{m+1}^{i_{m+1}} \cdots x_N^{i_N} \in$$

$$k((x_1, \ldots, x_m))(x_{m+1}, \ldots, x_N) .$$

iii) By using an argument as in (ii) we may conclude that

$$k((x_1,\dots,x_i)) = k((x_1,\dots,x_m))(x_{m+1},\dots,x_i), \quad m+1 \leq i \leq N.$$

Now, the W.P.T. applied to f_{i+1}, $m+1 \leq i \leq N$, give us

$$f_{i+1}(X_1,\dots,X_{i+1}) = V_i(X_1,\dots,X_{i+1})\, g_{i+1}((X_1,\dots,X_i),X_{i+1})$$

where V_i is a unit in $k((X_1,\dots,X_{i+1}))$ and g_{i+1} a monic polynomial in X_{i+1} with coefficients in $k((X_1,\dots,X_i))$. It follows that x_{i+1} is integral over $k((x_1,\dots,x_i)) = k((x_1,\dots,x_m))(x_{m+1},\dots,x_i)$ (since $g_{i+1}((x_1,\dots,x_i),x_{i+1}) = 0$). Hence, each x_i is integral over $k((x_1,\dots,x_m))$, $m+1 \leq i \leq N$.

iv) By (i) and (iii)

$$\operatorname{depth}(\underline{a}) = \dim(\square) = \dim(k((x_1,\dots,x_m))) = m.$$

Therefore $\operatorname{height}(\underline{a}) = N - \operatorname{depth}(\underline{a}) = N-m$.

2) If $m = \operatorname{depth}(\underline{a}) > 0$, $\underline{a}$ is clearly not M-primary. Conversely if $m=0$, we have $\underline{a} \cap k((X_1)) = (X_1^r)$, $r > 0$. Thus

$$\operatorname{depth}(\underline{a}) = \dim(\square) = \dim(k((x_1))) = \dim(k((X_1))/_{(X_1^r)}) = 0$$

implies that $\underline{a}$ is M-primary.

<u>Corollary 1.1.8.</u>- Let $\square$ be an irreducible algebroid curve over k. Then, if $N \geq \operatorname{Emb}(\square)$ there exists a prime ideal $\underline{p} \subset k((\underline{X}))$ such that $\square \cong k((\underline{X}))/_{\underline{p}}$ and if we set $x_i = X_i + \underline{p}$, the following properties hold:

1) $\underline{p} \cap k((X_1,\dots,X_i)) \neq (0)$, $2 \leq i \leq N$; $\underline{p} \cap k((X_1)) = (0)$.

2) There exist non zero (irreducible) series $f_i \in \underline{p} \cap k((X_1,\dots,X_i))$, $2 \leq i \leq N$, such that

$$\underline{v}(f_i(X_1,\dots,X_i)) = \underline{v}(f_i(0,\dots,0,X_i)).$$

3) x_1 is formally independent over k.

4) $k[[x_1,\dots,x_N]] = k[[x_1]][x_2,\dots,x_N]$

5) $k[[x_1,\dots,x_N]]$ is an integral extension of $k[[x_1]]$.

Remark 1.1.9.- The five above properties will be assumed in the sequel. Therefore we shall write $\square = k[[x_1]][x_2,\dots,x_N]$, x_1 being formally independent over k, and each x_i integral over $k[[x_1]]$, $2 \leqslant i \leqslant N$.

By (5) $\square$ is integral over $k[[x_1]]$, and hence algebraic over $k((x_1))$. Every $z \in \underline{m}$ is a zero of an irreducible polynomial over $k((x_1))$ having its coefficients in $k[[x_1]]$,

$$g((x_1),Z) = Z^s + b_{s-1}(x_1)\, Z^{s-1} + \dots + b_o(x_1).$$

This polynomial is actually distinguished, i.e., $b_j(0) = 0$ for $0 \leqslant j \leqslant s-1$. Indeed, if i is the smallest integer for which $b_i(0) \neq 0$, we may use the W.P.T., applied to the two variable series $g((X_1),Z)$, in order to find a new polynomial $g^*((X_1),Z)$ with coefficients in $k[[X_1]]$ and degree $i < s$ (note that $i > 0$ since g is not a unit) and a unit $U(X_1,Z) \in k[[X_1,Z]]$ such that

$$g((x_1),Z) = U(x_1,Z).\, g^*((x_1),Z) \quad .$$

Hence $g^*((x_1),z) = 0$ which is a contradiction, because $g((x_1),Z)$ was the polynomial with z as a zero which had the minimum degree.

$$\text{Since} \quad k[[X_1]][Z]/_{(g((X_1),Z))} \cong k[[X_1,Z]]/_{(g(X_1,Z))}$$

(Zariski-Samuel, [29], p. 146), the polynomial $g((X_1),Z)$ is also irreducible as two variable series.

Remark 1.1.10.- If $\text{Emb}(\square) \leqslant N$ the curve $\square$ can be embedded in an N-space. When $N=2$ the curve is said to be plane. In this case, for an embedding in a 2-space (=algebroid plane or plane if there is no confusion), the ideal $\underline{p}$ is actually principal, $\underline{p} = (f(X,Y))$. Furthermore if $X=X_1$ does not divide the leading form of f, the five properties in 1.1.8. are trivially satisfied.

Proposition 1.1.11.- $\square$ is a regular domain if and only if $\mathrm{Emb}(\square)$ is one.

This is a well known result in commutative algebra, but it can also be obtained from the normalization theorem as a corollary. Moreover, $\square$ is regular if and only if $\square$ is isomorphic as k-algebra to a formal power series ring in one indeterminate over k.

2. THE TANGENT CONE.

In this section we shall study the tangent cone of an irreducible algebroid curve from an algebraic and geometric view-point.

Let $\square$ be an irreducible algebroid curve over the algebraically closed field k. Consider the graded ring

$$gr_{\underline{m}}(\square) = \bigoplus_{n=o}^{\infty} \underline{m}^i/\underline{m}^{i+1} .$$

Definition 1.2.1.- The tangent cone to the curve $\square$ is defined to be the affine algebraic variety $\mathrm{Spec}(gr_{\underline{m}}(\square))$.

If a basis $\{x_i\}_{1\leqslant i\leqslant N}$ of $\underline{m}$ is given, the graded ring $gr_{\underline{m}}(\square)$ is generated as k-algebra by the classes $\{x_i+\underline{m}^2\}_{1\leqslant i\leqslant N}$. Then there is a canonical epimorphism

$$\begin{array}{ccc} k(X_1,\ldots,X_N) & \longrightarrow & gr_{\underline{m}}(\square) \\ X_i & \longmapsto & x_i+\underline{m}^2 \end{array}$$

($\{X_i\}_{1 \leq i \leq N}$ being indeterminates over k), and therefore an isomorphism

$$gr_{\underline{m}}(\square) \cong k[X_1,\ldots,X_N] / \underline{a}$$

where $\underline{a}$ is a homogeneous ideal of $k[X_1,\ldots,X_N]$. This gives rise to an embedding of the tangent cone in k^N : It is the affine algebraic variety in k^N defined by the ideal $\underline{a}$.

Remark 1.2.2.- If $\square = k((\underline{X}))/_{\underline{p}}$ is the curve defined by the ideal $\underline{p}$ and if $\underline{a}$ is the homogeneous ideal of $k[\underline{X}]$ defining its tangent cone as above, then $\underline{a}$ is generated by the leading forms of all the series in $\underline{p}$.

Proposition 1.2.3.- There is no series in $\underline{p}$ with a leading form of type aX_1^m, $a \in k$, $a \neq 0$.

Proof: Otherwise $X_1^m \in \underline{a}$, $m > 0$, we would have $X_1 \in \sqrt{\underline{a}}$. Now, as the series $f_2 \in \underline{p}$ (corollary 1.1.8.) has a leading form of type $bX_2^q + X_1 g(X_1,X_2)$, we would conclude that $X_2 \in \sqrt{\underline{a}}$. In the same way, by replacing f_i instead of f_2, $i > 2$, and using induction, we would obtain $X_i \in \sqrt{\underline{a}}$. Then,

$$\dim gr_{\underline{m}}(\square) = \dim {}^{k[X_1,\ldots,X_N]}/_{\underline{a}} = \dim {}^{k[X_1,\ldots,X_N]}/_{(X_1,\ldots,X_N)} = 0$$

would be a contradiction, since $\dim gr_{\underline{m}}(\square) = 1$ (see Zariski-Samuel, [29], p. 235).

Corollary 1.2.4.- Let $g((x_1),Z) = Z^s + b_{s-1}(x_1)Z^{s-1} + \ldots + b_o(x_1)$ $b_j(x_1) \in k[[x_1]]$, the irreducible polynomial over $k((x_1))$ of an element $z \in m$. Consider g as a two variable series. Then, the leading form of g is a power of a linear form and $\underline{\upsilon}(g((x_1),Z)) = \underline{\upsilon}(g((0),Z)) = s$. In particular, the series f_i in 1.1.8. may be taken to be $g((X_1),X_i)$.

Proof: First, we prove that if a series $f(X,Y)$ is irreducible then its leading form $f_r(X,Y)$ is a power of a linear form. In fact, making a linear change of variables, f may be considered to be regular in Y of order r and hence, by the W.P.T., we may assume that it is a polynomial of $k[[X]][Y]$ of degree r. Then $f'(X,Y')=f(X,XY')/_{Y'^r}$ is an irreducible monic polynomial of $k[[X]][Y']$, and hence by the Hensel's lemma $f_r(1,Y')=f'(0,Y')=(Y'+a)^r$ with $a\in k$, and so $f_r(X,Y)=(Y+aX)^r$. Now, $g((x_1),Z)$ is irreducible as two variable series (see 1.1.9.) and so its leading form is $(ax_1+bZ)^r$ with $a,b\in k$. By the previous proposition $b\neq 0$, and since g is distinguished we have $r=s$. Hence the proof follows easily.

Lemma 1.2.5.- The tangent cone to a curve is a straight line.

Proof: Choose the series $g_i((X_1),X_i)$ in the above corollary instead of $f_i(X_1,\ldots,X_i)$. The leading form of g_i is of type $(X_i+\alpha_i X_1)^{r_i}$. Then, since $\dim k[X_1,\ldots,X_N]/\sqrt{\underline{a}}=1$, we must have

$$\sqrt{\underline{a}} = (X_2+\alpha_2 X_1,\ldots, X_N+\alpha_N X_1).$$

It follows that the tangent cone is the straight line defined by

$$X_2+\alpha_2 X_1 = \ldots = X_N+\alpha_N X_1 = 0.$$

3. LOCAL PARAMETRIZATION.

Let $\square$ be an irreducible algebroid curve over k. Let F be the quotient field of $\square$. Choose a normalized basis $\{x_i\}_{1\leqslant i\leqslant N}$ (i.e., a basis for which the conditions of 1.1.8. hold) of the maximal ideal $\underline{m}$.

Since $k((x_1))[x_2,\ldots,x_N]$ is a subfield of F containing $\square = k[[x_1]][x_2,\ldots,x_N]$, we have:

$$F = k((x_1))(x_2,\ldots,x_N),$$

and therefore the field extension $F/_{k((x_1))}$ is finite.

On the other hand, as x_1 is formally independent over k, the ring $k[[x_1]]$ is a complete discrete valuation ring of $k((x_1))$. Furthermore, its relative integral closure in F is the same as the relative closure of $\square$, i.e.

$$\overline{k[[x_1]]} = \overline{\square}.$$

Theorem 1.3.1.- $\overline{\square}$ is a complete discrete valuation ring of F which dominates $k[[x_1]]$.

If $\bar{m}$ is the maximal ideal of $\overline{\square}$, $t \in \bar{m}-\bar{m}^2$, and T is an indeterminate over k, the homomorphism given by

$$h : k[[T]] \longrightarrow \overline{\square}$$
$$T \longmapsto t$$

is a k-isomorphism.

Proof: Since $F/_{k((x_1))}$ is finite and $k((x_1))$ is complete for the associated valuation to $k[[x_1]]$, there is only one valuation ring A of F which lies over $k[[x_1]]$. As $\overline{\square} = \overline{k[[x_1]]}$ is the intersection of all the valuation rings lying over $k[[x_1]]$, it is trivial that $\overline{\square} = A$. Finally, the valuation ring is discrete since $\overline{\square}$ lies over $k[[x_1]]$, which is a discrete valuation ring itself.

In order to prove the second statement, first we shall prove that $\overline{\square}$ has k as a coefficient field.

Let $\underline{q}$ be the maximal ideal of $k[[x_1]]$. We may construct an injective field homomorphism

$$k \equiv k[[x_1]]/_{\underline{q}} \longrightarrow \overline{\square}/_{\bar{m}}$$

which is also onto, since $\overline{\square} / _{\overline{m}}$ is algebraic over k and k is algebraically closed.

The homomorphism h is well defined, because $t \in \overline{m}$ and the topology over $\square$ induced by its valuation is actually the $\overline{m}$-adic topology of $\overline{\square}$.

The kernel of h will be an ideal of $k[[T]]$, hence it will be generated by T^m for some m. We claim that $m = 0$. In fact, if $m > 0$, we would have $t^m = 0$, thus $t = 0$, and we would get a contradiction. Actually h is an epimorphism, since t is a uniformizing parameter for the complete discrete valuation ring $\overline{\square}$.

<u>Definitions 1.3.2.-</u> We shall say that a uniformizing parameter t for the discrete valuation ring $\overline{\square}$ is a <u>uniformizing parameter for the curve $\square$</u>. Henceforth, for a such t, we shall write

$$\overline{\square} = k[[t]] \quad \text{and} \quad F = k((t)).$$

Let $\underline{v}$ be the normalized natural valuation of $k((t))$. We shall say that $\underline{v}$ is the <u>associated valuation to $\square$</u>. If $z \in \overline{\square}$, $z = s(t)$ with $s(T) \in k[[T]]$, we have

$$\underline{v}(z) = \underline{\upsilon}(s(T)).$$

The uniformizing parameters are exactly those elements in F having value 1 in $\underline{v}$.

<u>Remark 1.3.3.</u>- The natural valuation of $\square$ does not depend on x_1. Thus it is intrinsic of $\square$ and it may be defined using any uniformizing parameter t for $\overline{\square}$.

<u>Proposition 1.3.4.</u>- The ramification index of $\overline{\square}$ over $k((x_1))$ is the degree of the extension $F / _{k((x_1))}$.

<u>Proof</u>: Let e be the ramification index of $\overline{\square}$ over $k((x_1))$. Denote by $\underline{v}$ the associated valuation to $\square$. If $\underline{v}(x_1) = h$, and if $s(x_1)$ is any formal power series in x_1, we have $\underline{v}(s(x_1)) = h\ \underline{\upsilon}(s)$, which implies that

$$e = (Z : hZ) = h.$$

Now, if t denotes a uniformizing parameter for $\square$, we have $x_1 = x_1(t)$, with $\underline{\upsilon}(x_1) = e$. Using successive divisions in $k[[t]]$ we may check the formula:

$$(1) \qquad \overline{\square} = \bigoplus_{i=0}^{e-1} k[[x_1]]\ t^i$$

This equality shows that $\{1, t, \ldots, t^{e-1}\}$ is a basis of the vector space F over $k((x_1))$. It follows that $[F : k((x_1))] = e = \underline{v}(x_1)$.

Note that (1) also proves that $\overline{\square}$ is a $k[[x_1]]$-module free of rank e.

<u>Proposition 1.3.5.-</u> For any $z \in \overline{\square}$, $z \neq 0$, and $\underline{v}(z) \geqslant 1$, the following statements hold:

i) z is formally independent over k.

ii) $\underline{v}$ is an extension of the natural valuation of $k[[z]]$ with ramification index $\underline{v}(z)$.

iii) $\overline{\square} / _{k[[z]]}$ is an integral extension, and $\overline{\square}$ is a $k[[z]]$ - module free of rank $\underline{v}(z)$.

iv) $k((t)) / _{k((z))}$ is an algebraic extension of degree $\underline{v}(z)$.

<u>Proof:</u> i) The element t is formally independent over k, and $z \in k[[t]]$ with $\underline{v}(z) \geqslant 1$, thus z is formally independent over k.

ii) The maximal ideal of $k[[z]]$ is trivially contained in the maximal ideal of $\overline{\square}$, then $\underline{v}$ is an extension of the natural valuation of $k[[z]]$.

To prove that the ramification index of $\overline{\square}$ over $k((z))$ is $\underline{v}(z)$, use a formula as (1) in 1.3.4.:

$$\overline{\square} = \bigoplus_{i=o}^{\underline{v}(z)-1} k((z))\, t^i$$

iii) By the formula in ii), $\overline{\square} = k((z))(t)$ and $\overline{\square}$ is a free $k((z))$-module of rank $\underline{v}(z)$. We must prove that t is integral over $k((z))$.

If $z = s(t)$, the two variable series $S(Z,T) = Z - s(t)$ is regular in T of order $\underline{v}(z)$. By the W.P.T., there exist a unit $U(Z,T) \in k((Z,T))$ and a monic polynomial $P((Z),T)$ of degree $\underline{v}(z)$ with coefficients in $k((Z))$ such that:

$$S(Z,T) = U(Z,T).P((Z),T).$$

Since $U(z,t) \neq 0$, we may deduce that $P((z),t) = 0$, and thus t is integral over $k((z))$.

iv) $F = k((z))(t)$, and t is a zero of an irreducible polynomial over $k((z))$ of degree $\underline{v}(z)$. The proof is evident.

<u>Corollary 1.3.6.</u>- Let $\square$ be a curve and $\{y_i\}_{1 \leq i \leq N}$ any basis of its maximal ideal, with $y_1 \neq 0$. Then, the following statements hold:

i) y_1 is formally independent over k.

ii) The extension $\square / k((y_1))$ is integral.

iii) $\square = k((y_1))(y_2, \ldots, y_N)$.

(i.e., properties (3), (4), and (5) in the corollary 1.1.8. hold.)

<u>Proof:</u> (i) and (ii) follow from the above proposition. We shall prove (iii). For each i, $2 \leq i \leq N$, the irreducible polynomial g of y_i over $k((y_1))$ is distinguished and it has its coefficients in $k((y_1))$, so

$$k((Y_1, Y_i)) /_{(g)} \cong k((Y_1))(Y_i) /_{(g)}$$

(see Zariski-Samuel, [29], p. 146). Then $k[[y_1, y_i]] = k[[y_1]][y_i]$.

Now, we get $k[[y_1, \dots, y_i]] = k[[y_1]][y_2, \dots, y_i]$ by using induction on i:

$$k[[y_1, \dots, y_i]] = k[[y_1, \dots, y_{i-1}]][[y_i]] = k[[y_1, y_i]][y_2, \dots, y_{i-1}]$$
$$= k[[y_1]][y_2, \dots, y_i].$$

<u>Proposition 1.3.7</u>.- Let $L \supset F$ be a finite extension of the quotient field F of a curve $\square$. Then, the integral closure $\overline{\square}_L$ of $\square$ in L is a complete discrete valuation ring of L. If u is a uniformizing parameter for $\overline{\square}_L$ then $\overline{\square}_L = k[[u]]$, $L = k((u))$, and $\underline{v}_u(t) = [L : F]$. In particular, any finite extension of a series field $k((t))$ is of type $k((u))$.

<u>Proof</u>: It is similar to the proof of theorem 1.3.1., replacing F by L, $\overline{\square}$ by $\overline{\square}_L$ and x_1 by t. The equality $\underline{v}_u(t) = [F : L]$ follows from the last corollary.

Let $\square$ be an irreducible curve over k and $L = k((u))$ a finite extension of its quotient field F. If $\{y_i\}_{1 \leqslant i \leqslant N}$ denotes a basis of the maximal ideal $\underline{m}$ of $\square$, each y_i is a formal power series in u of positive order ; i.e., we have :

$$(1) \qquad y_i = y_i(u) \quad , \quad 1 \leqslant j \leqslant N, \quad \text{with} \quad \underline{v}(y_i) > 0.$$

<u>Definition 1.3.8</u>.- A set as (1) will be called <u>a local parametric representation of the curve $\square$ in the considered basis</u>. Moreover, we shall say that (1) are <u>local parametric equations for the curve</u>.

<u>Remark 1.3.9</u>.- Let $\{y_i\}_{1 \leqslant i \leqslant N}$ be a set of elements of $\square$, and denote by $\{a_i\}_{1 \leqslant i \leqslant N} \subset k$ its residues modulo $\underline{m}$. We have:

$$(2) \qquad y_i = a_i + x_i(u) \quad , \quad 1 \leqslant i \leqslant N,$$

with $x_i(u) \in \underline{m}$. Let us assume that $\{x_i\}_{1\leq i\leq N}$ is a basis of $\underline{m}$. The set (2) is also called local parametrization of the curve $\square$, and the point $(a_1,\ldots,a_N)\in k^N$ is said to be the <u>centre</u> of the parametrization.

(1) is a local parametric representation with centre in $\underline{O}=(0,\ldots,0)\in k^N$. We shall assume henceforth that the centre is $\underline{O}$ unless otherwise stated.

<u>Definition 1.3.10.</u>- We say that a local parametric representation (1) is <u>primitive</u> when $L = F$.

<u>Remark 1.3.11.</u>- Let $y_i = y_i(t)$, $1\leq i\leq N$, be a primitive parametric representation of a curve $\square$ in the basis $\{y_i\}_{1\leq i\leq N}$ of its maximal ideal $\underline{m}$. Then any other parametric representation in that basis is obtained by a substitution of type

$$t = c_h t^h + c_{h+1} t^{h+1} + \ldots \quad , \quad h>0, \quad c_i \in k \ , \quad c_h \neq 0.$$

The new representation is primitive if and only if $h=1$.

<u>Remark 1.3.12.</u>- Given any equalities

$$y_i = y_i(u) \in k((u)) \quad , \quad \underline{v}(y_i) > 0 \ , \quad 1\leq i\leq N,$$

there is always a curve $\square$ which has those equalities as a local parametric representation with centre in $\underline{O}$. In fact, if we consider the ring homomorphism

$$\begin{array}{rcll} f : k((\underline{X})) & \longrightarrow & k((u)) & , \qquad \underline{X} = \{X_i\}_{1\leq i\leq N}, \\ X_i & \longmapsto & y_i(u) & , \end{array}$$

the kernel of f is a prime ideal $\underline{p}\subset k((\underline{X}))$, and trivially our statement holds for the curve $\square = k((\underline{X}))/\underline{p}$.

Definitions 1.3.13.- An element $z \in \square$, $z \neq 0$, is called a parameter of the curve when the principal ideal $\square z$ is $\underline{m}$-primary. Since $\square$ is a domain with only two prime ideals (0) and $\underline{m}$, each non zero element of $\underline{m}$ is a parameter.

Let x be a parameter of $\square$. We define the separability degree (resp. inseparability index) of x to be the separability degree (resp. inseparability index) of the extension $F / k((x))$. We shall denote these numbers by $n_s(x)$ and $i(x)$ respectively.

If the characteristic of k is $p > 0$, we always have:

$$n_s(x).p^{i(x)} = \left[F : k((x)) \right] = \underline{v}(x).$$

If the characteristic is zero, the above equality holds with 1 instead of $p^{i(x)}$.

The parameter x is said to be separable iff $i(x) = 0$, and inseparable otherwise.

Proposition 1.3.14.- Let k be a field of characteristic $p > 0$, x a parameter of a curve $\square$, and t a uniformizing parameter of $\overline{\square}$. Then:

i) The inseparability index of x is the maximum integer i for which $x \in \overline{\square}^{p^i} = k[[t^{p^i}]]$.

ii) x is separable if and only if $(dx/dt) \neq 0$.

Proof: The degree of the extension $F / k((x))$ is $n = \underline{v}(x)$, and t is a primitive element of this extension; then the inseparability index of x is the inseparability index of t over $k((x))$.

Put $x = s(t^{p^i})$. By the W.P.T. applied to the series $X - S(T)$, there exists a unit $U(X,T)$ in $k[[X,T]]$ and a monic polynomial $P(X,T)$ in $k[[X]][T]$ of degree n/p^i sucht that

$$U(X,T)\,(X - S(T)) = P(X,T).$$

P is irreducible over $k((x))$, and $P(x,t^{p^i}) = 0$, then obviously this must be the minimal polynomial of t over $k((x))$.

On the other hand,

$$\frac{\partial P}{\partial T}(x,t^{p^i}) = - U(x,t^{p^i}) \frac{dS}{dT}(t^{p^i}) \neq 0,$$

and hence i is exactly the inseparability index of t over $k((x))$. This completes the proof of (i). The part (ii) follows trivially from (i).

<u>Proposition 1.3.15.</u>- Let $\square$ be a curve with quotient field F. Assume that $\{y_i\}_{1 \leqslant i \leqslant N}$ is a basis of the maximal ideal $\underline{m}$, $y_1 \neq 0$. Then, there exists another basis $\{y_i'\}_{1 \leqslant i \leqslant N}$, with $y_1' = y_1$, obtained from the first by an inversible linear change with coefficients in k, such that each y_i', $2 \leqslant i \leqslant N$, is a primitive element of the extension $F / k((y_1))$.

<u>Proof:</u> We shall distinguish two cases, according to whether y_1 is a separable parameter or not.

First, let y_1 be separable, and write $L = k((y_1))$. We have $F = L(y_2, \ldots, y_N)$ and each y_i is separable over L. We use induction on N. For $N=2$, it is trivial. Assume $N > 2$ and set $F' = L(y_2, \ldots, y_{N-1})$. By the induction hypothesis there is an inversible linear change of variables $y_2, \ldots, y_{N-1}$ with coefficients in k, such that the new elements $y_2'', \ldots, y_{N-1}''$ are primitive elements of F'/L.

Denote by $f_1, \ldots, f_n$ ($n = [F : L]$) the L-isomorphisms from F into an algebraic closure of F. Since k is infinite and $F = L(y_2'', y_N)$ there exists $a \in k$ such that

$$f_i(a y_2'' + y_N) \neq f_j(a y_2'' + y_N) \quad \text{for } i \neq j .$$

Thus $y_N' = a y_2'' + y_N$, for a such a, is a primitive element of F / L.

Now, in the same way, we may substitute y_i'' by $y_i' = y_i'' + a_i y_N'$, $2 \leq i \leq N-1$, choosing an appropriate $a_i \in k$ to be y_i' a primitive element.

In order to prove the second case, assume that y_1 is inseparable and set $p = \text{charact. } k > 0$, $n = [F : k((y_1))]$, $i = i(y_1)$, $n' = n_s(y_1)$. Denote by $k_i = k((\tau))$ the maximum purely inseparable extension of $k((y_1))$ contained in F. F/k_i is thus separable and we have:

$$[k_i : k((y_1))] = p^i .$$

Then for the curve with local ring $\square^* = k[[\tau, y_2, \ldots, y_N]]$, the conditions in the above case hold, since τ is a separable parameter of $\square^*$. Therefore $y_2, \ldots, y_N$ may be substituted by $y_2', \ldots, y_N'$, by means of an inversible linear change with coefficients in k, such that each y_i', $2 \leq i \leq N$, is a primitive element of F/k_i.

Moreover, the y_i' may be assumed to be separable: In fact, since a parametric representation with a uniformizing as parameter is primitive, by 1.3.14. it must exist a y_j' which is so. Hence, since k is infinite, there exist values of b in k such that, for $i \neq j$, $y_i' + b y_j'$ is separable and a primitive element of F/k_i.

Now, we shall prove that y_i' is a primitive element of $F/k((y_1))$. The separability degree of y_i' over $k((y_1))$ is n', then $k((y_1))(y_i')$ contains the maximum separable extension k_s of $k((y_1))$ between $k((y_1))$ and F. Let h be the minimum integer for which $y_i'^{p^h} \in k_s = k[[t^{p^i}]]$. Since y_i' is a separable parameter, it is necessary that $h \geq i$, and actually $h = i$. Then, the degree of $k((y_1))(y_i')$ over $k((y_1))$ is $n'.p^i = [F : k((y_1))]$, and hence y_i' is a primitive element of $F/k((y_1))$. This completes the proof of the proposition.

4. TRANSVERSAL PARAMETERS. MULTIPLICITY.

Henceforth we shall work, as we have pointed out, with bases of the maximal ideal $\underline{m}$ having the normalization properties of 1.1.8. These bases were called normalized bases. We have already seen that to obtain those properties it suffices to impose some conditions to x_1. In this section we shall give several algebro-geometric characterizations of these parameters, which will be called transversal parameters. We shall also compute the multiplicity of the curve by using transversal parameters.

Definition 1.4.1.- Let $\square$ be an irreducible algebroid curve. We say that a parameter $x \in \underline{m}$ is transversal if $x + \underline{m}^2$ is not nilpotent in $gr_{\underline{m}}(\square)$.

Remark 1.4.2.- For a geometric interpretation of the concept of transversal parameter we shall use an embedding in k^N by means of the isomorphism

$$gr_{\underline{m}}(\square) \cong k(X_1, \ldots, X_N) / \underline{a}$$

defined by a basis $\{x_i\}_{1 \leq i \leq N}$ of $\underline{m}$, which needs not be minimal.

If $x \in \underline{m} - \underline{m}^2$, the equation

$$(1) \qquad x + \underline{m}^2 = \sum_{i=1}^{N} \lambda_i (x_i + \underline{m}^2)$$

has some solutions $\lambda_i \in k$, $1 \leq i \leq N$.

Suppose that Ω_x is the linear variety in k^N given by the linear forms:

$$\lambda_1 X_1 + \ldots + \lambda_N X_N = 0$$

where $(\lambda_1, \ldots, \lambda_N)$ ranges over the set of solutions of (1).

The dimension of Ω_x is $r-1$, being $r = \mathrm{Emb}(\square)$. When $N = r$, i.e., when the basis is minimal, Ω_x is a hyperplane.

Theorem 1.4.3.- Let x a parameter of the curve $\square$. The following statements are equivalent:

a) x is a transversal parameter.

b) $x \notin \underline{m}^2$ and for any embedding, the linear variety Ω_x does not contain the tangent cone.

c) Every basis $\{x_i\}_{1 \leq i \leq N}$ of $\underline{m}$, with $x = x_1$, is normalized.

d) There exists a basis $\{x_i\}_{1 \leq i \leq N}$ of $\underline{m}$, with $x = x_1$, which is normalized.

e) x is an element of $\underline{m}$ having the minimum value in $\underline{v}$.

Proof: (a) $\Longleftrightarrow$ (b) Suppose $x + \underline{m}^2 = (\sum_{i=1}^{N} \lambda_i x_i) + \underline{m}^2$ and let $\underline{a}$ the ideal which defines the tangent cone. The element $x + \underline{m}^2$ is not nilpotent in $\mathrm{gr}_{\underline{m}}(\square)$ if and only if $\sum_{i=1}^{N} \lambda_i X_i \notin \sqrt{\underline{a}}$. In other words, x is a transversal parameter if and only if $x \notin \underline{m}^2$ and Ω_x does not contain the tangent cone.

(b) $\Longrightarrow$ (c) Let $\{x_i\}_{1 \leq i \leq N}$ be a basis of $\underline{m}$ with $x = x_1$. By 1.3.6. the three latter normalization properties are true. However we must prove the former two.

If $\underline{p} \cap k[[X_1]] \neq (0)$, then $X_1 \in \underline{p}$ since $\underline{p}$ is prime. Hence $x = x_1 = 0$ which is a contradiction.

Now, we take for each i, $2 \leq i \leq N$, the irreducible polynomial of x_i over $k((x_1))$. Its leading form as two variable series is of type $(a X_i + b X_1)^s$; therefore, the hyperplane $a X_i + b X_1 = 0$ contains the tangent cone. But since (b) is true, we must have $a \neq 0$. Thus:

$$\underline{\upsilon}\,(g((0),X_i)) = \underline{\upsilon}\,(g((X_1),X_i)).$$

(c) $\Longrightarrow$ (d) It is evident.

(d) $\Longrightarrow$ (e) Take a basis $\{x_i\}_{1\leqslant i\leqslant N}$ normalized with $x=x_1$. It suffices to prove $\underline{v}(x_1) \leqslant \underline{v}(x_i)$, $2 \leqslant i \leqslant N$.

The irreducible polynomial of x_i over $k((x_1))$ verifies $\underline{\upsilon}(g((0),X_i)) = \underline{\upsilon}(g((X_1),X_i)$ (see 1.2.4.). Since g is monic in X_i, the equality

$$g((x_1),x_i) = 0$$

implies $\underline{v}(x_1) \leqslant \underline{v}(x_i)$.

(e) $\Longrightarrow$ (b) First $x \notin \underline{m}^2$ because $\underline{v}(x)$ is the minimum value for $\underline{v}$ in $\underline{m}$. Therefore, for any embedding we can make a linear change of bases in $\underline{m}$ to have $x - x_1 \in \underline{m}^2$. Then no series in the ideal $\underline{p}$ defining the curve can have a leading form of type $a X_1^m$, $a \neq 0$; or the same, the tangent cone is not contained in the hyperplane $X_1 = 0$, which is one of the hyperplanes defining Ω_x.

Remark 1.4.4.- For algebroid plane curves (Emb($\square$) $\leqslant 2$), the condition in the definition of transversal parameter can be substituted by a stronger condition:

" x is transversal if and only if $x + \underline{m}^2$ is not a zero divisor in $gr_{\underline{m}}(\square)$ ".

The equivalence is obvious, because in this case the ideal $\underline{a}$ verifying $gr_{\underline{m}}(\square) \cong k[X_1,X_2]/_{\underline{a}}$ is primary; then, any zero divisor in this graded ring is nilpotent.

However, for twisted curves, $x + \underline{m}^2$ may be a zero divisor without being nilpotent in $gr_{\underline{m}}(\square)$. For instance, for the curve given by the parametric representation:

$$x = t^4$$
$$y = t^5$$
$$z = t^{10} + t^{11} \quad ,$$

over the complex field, $x+\underline{m}^2$ is not nilpotent in the graded ring since x has the minimum value in $\underline{v}$, but it is a zero divisor since the following equality holds

$$(x + \underline{m}^2)(z + \underline{m}^2) = (y^2 x + y^3) + \underline{m}^3 = 0 \in gr_{\underline{m}}(\square),$$

with $z + \underline{m}^2 \neq 0$.

<u>Definitions 1.4.5.</u>- The <u>multiplicity of the curve</u>, $e(\square)$, is defined to be the multiplicity of the maximal ideal $\underline{m}$ of $\square$; i.e., it is the integer e for which $\dim_k(\square/_{\underline{m}}n) = en + q$ for n large enough.

A <u>superficial element of order 1</u> for the curve $\square$ is a element $x \in \underline{m}$ such that there exists an integer $c > 0$ for which

$$(\underline{m}^n : \square x) \cap \underline{m}^c = \underline{m}^{n-1}$$

for n large enough. (See Zariski-Samuel, [29] , p. 285-294.)

<u>Proposition 1.4.6.</u>- Any transversal parameter x of a curve $\square$ is a superficial element of order 1 for $\square$.

<u>Proof</u>: We must prove the formula in the above definition of superficial element for some c . The containement $\supset$ is always true, thus it suffices to prove the other. We shall distinguish two cases, according as $x + \underline{m}^2$ be it a zero divisor in $gr_{\underline{m}}(\square)$ or not.

<u>1st case: $x + m^2$ is not a zero divisor.</u>

Take $c = 1$. Let $z \in \underline{m}$ be an element such that $xz \in \underline{m}^n$ but $z \notin \underline{m}^{n-1}$. Denote by s the minimun integer for which $z \in \underline{m}^s - \underline{m}^{s+1}$ $(s+1 \leqslant n-1)$. Then $z + \underline{m}^{s+1} \neq 0$, and

$$(z + \underline{m}^{s+1})\,(x + \underline{m}^2) = zx + \underline{m}^{s+2}$$

Since $s+2 \leqslant n$ the right hand side member is 0, which is a contradiction.

<u>2nd case: $x + m^2$ is a zero divisor.</u>

Let us consider a basis $\{x_i\}_{1 \leqslant i \leqslant N}$ of $\underline{m}$ with $x = x_1$. Set $\bar{x}_i = x_i + \underline{m}^2$, $1 \leqslant i \leqslant N$. Let $\underline{a}$ be the ideal of $k[X_1, \ldots, X_N]$ which defines the tangent cone. Since

$$\sqrt{\underline{a}} = (X_2 + \alpha_2 X_1, \ldots, X_N + \alpha_N X_1)$$

(see 1.2.5.), $\sqrt{\underline{a}}$ is the unique isolated prime divisor of $\underline{a}$. Then, another associated prime divisor of $\underline{a}$ is an embedded component. But $X_1 \notin \sqrt{\underline{a}}$ since x_1 is transversal, and as $\bar{x}_1$ is zero divisor in $gr_{\underline{m}}(\square)$, X_1 must belong to an embedded component. As

$$(X_2 + \alpha_2 X_1, \ldots, X_N + \alpha_N X_1, X_1) = (X_1, \ldots, X_N)$$

the ideal $(X_1, \ldots, X_N)$ is an embedded component. Furthermore, $(X_1, \ldots, X_N)$ is the unique embedded component of $\underline{a}$ because $\underline{a}$ is a homogeneous ideal, and therofore all its associated prime divisors are homogeneous.

Thus, any primary decomposition of $\underline{a}$ is of type

$$\underline{a} = \underline{q}_1' \cap \underline{q}_2',$$

where $\sqrt{\underline{q}_1'} = \sqrt{\underline{a}}$ and $\sqrt{\underline{q}_2'} = (X_1, \ldots, X_N)$. It follows that any primary decomposition of the ideal (0) in $gr_{\underline{m}}(\square)$ is of type

$$(0) = \bar{\underline{q}}_1 \cap \bar{\underline{q}}_2,$$

where $\sqrt{\bar{\underline{q}}_1} = (\bar{x}_2 + \alpha_2 \bar{x}_1, \ldots, \bar{x}_N + \alpha_N \bar{x}_1)$ and $\sqrt{\bar{\underline{q}}_2} = (\bar{x}_1, \ldots, \bar{x}_N)$.

Take an integer $c > 0$ for which $(\bar{x}_1, \ldots, \bar{x}_N)^c \subset \bar{q}_2$. Let $z \in \underline{m}^c$ be an element such that $xz \in \underline{m}^n$ but $z \notin \underline{m}^{n-1}$. Denote by s, as above, the integer such that $z \in \underline{m}^s - \underline{m}^{s+1}$ $(c+1 \leqslant s+1 \leqslant n-1)$. We have

$$(x_1 + m^2).(z + \underline{m}^{s+1}) = xz + \underline{m}^{s+2} = 0.$$

Since $\bar{x}_1 \notin \sqrt{\bar{q}_1}$, this relation implies $z + \underline{m}^{s+1} \in \bar{q}_1$. On the other hand, $z = h(x_1, \ldots, x_N)$, where h is a series of order $s \geqslant c$. Then

$$z + \underline{m}^{s+1} = h(x_1, \ldots, x_N) \in (x_1, \ldots, x_N)^c \subset \bar{q}_2 .$$

We can conclude that $z + \underline{m}^{s+1} \in \bar{q}_1 \cap \bar{q}_2 = (0)$, which is a contradiction since $z \notin \underline{m}^{s+1}$.

<u>Theorem 1.4.7.</u>- If x is a transversal parameter of a curve $\square$, then:

$$e(\square) = \left[F : k((x)) \right] = \underline{v}(x).$$

<u>Proof:</u> Since x is superficial of order 1 for $\square$, we have $e(\underline{m}) = e(\square x)$ (Zariski-Samuel, [29], page 294).

On the other hand $\square$ is a finite type $k[[x]]$-module, and thus we have (Zariski-Samuel, [29], page 300):

$$e(\square) = e(\square x) = \left[\square : k[[x]] \right] . \left[\square /_{\underline{m}} : k \right] = \left[\square : k[[x]] \right].$$

It only remains for us to compute $\left[\square : k[[x]] \right]$. Recall that this is the maximum number of $k[[x]]$-lineary independent elements in the module $\square$. By 1.2.15. there exists an element $y \in \underline{m}$ which is primitive for the extension $F / k((x))$. If we set $n = \left[F : k((x)) \right]$, the elements $\{1, y, \ldots, y^{n-1}\}$ in $\square$ are lineary independent over $k[[x]]$. Then

$$n \leq [\square : k((x))] \leq [\overline{\square} : k((x))] \leq n.$$

Corollary 1.4.8.- For a curve, the conditions:

(a) $e(\square) = 1$.

(b) $\square$ is integrally closed (and thus normal).

(c) There exists $x \in \underline{m}$ such that $\underline{v}(x) = 1$.

(d) $\mathrm{Emb}(\square) = 1$.

(e) $\square$ is regular.

are equivalent.

5. RESOLUTION OF SINGULARITIES.

Let $\square$ be the local ring of an irreducible algebroid curve over k. We shall denote by F its quotient field, and by $\underline{m}$ its maximal ideal.

Notations 1.5.1.- For each $x \in \underline{m}$, $x \neq 0$, let $x^{-1}\underline{m}$ be the set of quotients belonging to F which have the form z/x with $z \in \underline{m}$, and $\square_x = \square[x^{-1}\underline{m}]$ the $\square$-subalgebra of F generated by $x^{-1}\underline{m}$.

Let $T' = \bigcup_{x \in \underline{m}-\{0\}} \mathrm{Spec}(\square_x)$, and design by Ω_x, Ω'_x, etc... elements of $\mathrm{Spec}(\square_x)$.

Notice that since F is a field, all the quotients rings which we consider can be assumed to be subrings of F.

In T' we give the following equivalence relation:

$$\Omega_x \in \mathrm{Spec}(\square_x) \quad , \qquad \Omega_y \in \mathrm{Spec}(\square_y) \, ,$$

$$\Omega_x \sim \Omega_y \iff (\square_x)_{\Omega_x} = (\square_y)_{\Omega_y}$$

Denote finally by $\square$ the set $T'/\sim$.

The results which will be proved next are technical and they aim to study the set T.

First, for $y,z \in \underline{m} - \{0\}$, let $N_{y,z}$ be the multiplicatively closed set formed by the elements of type $(z/y)^r$, where $r \geq 0$. We shall denote by $\square_{y,z}$ the quotient ring $N_{y,z}^{-1} . \square_y$. The following equalities are evident:

i) $\square_{y,z} = \square_y \, [y/z]$.

ii) $\square_{y,z} = \square_{z,y}$.

<u>Proposition 1.5.2.</u>- Let x be a transversal parameter for $\square$, and $y \in \underline{m}$, $y \neq 0$. Then x/y is a unit in $\square_y$.

<u>Proof:</u> First, x is integral over $k[[y]]$, thus it is a zero of a monic polynomial, irreducible over $k((y))$ with coefficients in $k[[y]]$,

$$f(X,y) = X^k + A_{k-1}(y)\, X^{k-1} + \dots + A_o(y).$$

The same, y is integral over $k[[x]]$, thus it is a zero of a new monic polynomial, irreducible over $k((x))$ with coefficients in $k[[x]]$,

$$g(x,Y) = Y^h + B_{h-1}(x)\, Y^{h-1} + \dots + B_o(x).$$

We are going to find a relation between f and g. Since these irreducible polynomials are distinguished, they are also irreducible as two variable series. Moreover by the W.P.T. they differ only in a unit, and hence $\underline{\upsilon}(A_o(y)) = h$.

To look at this in another way, the leading form of f is of type $(a\,X + b\,Y)^r$. Since there always exists a basis of the maximal ideal $\underline{m}$ which contains x and y, and this basis is normalized because x is transversal, we must have $b \neq 0$ (see 1.2.3.). Therefore, $\underline{\nu}(A_o(y)) = h = r$ and the order of f is h.

Set $x' = x/y$. We have $0 = f(x,y) = f(x'y,y) = y^h f^*(x')$, where $f^*(x') = y^{k-h} x'^k + A^*_{k-1}(y)x'^{k-1} + \ldots + A^*_o(y)$ and $A^*_j(y) = A_j(y)/y^{h-j} \in k((y))$. Since y is not a zero divisor in $\square_y$ it follows that $f^*(x')=0$, and hence, since $A^*_o(y)$ is a unit in $k((y))$ (because the order of $A_o(y)$ is h), we have $A^*_o(y)^{-1} f^*(x')=0$. Collecting in the same side the terms of the last equality containing x', we obtain $x'f'(x') = 1$, where the element $f'(x') = -A^*_o(y)^{-1}(y^{k-h}x'^{k-1} + A^*_{k-1}(y)x'^{k-2} + \ldots + A^*_1(y))$ belongs to $\square_y$. Then x' is a unit in $\square_y$ as desired.

Corollary 1.5.3.- For $x, y \in \underline{m} - \{0\}$, where x is transversal, we have $\square_y = \square_{x,y}$.

Proposition 1.5.4.- If x is transversal, the map

$$\begin{array}{rcl} \psi_x : \mathrm{Spec}(\square_x) & \longrightarrow & T \\ \Omega_x & \longmapsto & \Omega_x\sim \end{array}$$

is bijective.

Proof: It is injective because $\Omega_x \sim \Omega'_x \iff (\square_x)_{\Omega_x} = (\square_x)_{\Omega'_x} \iff \Omega_x = \Omega'_x$.

In order to prove that it is surjective, take $\Omega_y \in \mathrm{Spec}(\square_y)$, $y \neq 0$, $y \in \underline{m}$. Then the contracted ideal of Ω_y by the canonical homomorphism

$$\square_x \longrightarrow \square_{x,y} = \square_y$$

is a prime ideal Ω_x of $\square_x$ which does not meet $N_{x,y}$ and therefore verifies

$$(\square_x)_{\Omega_x} = (\square_y)_{\Omega_y} \Longrightarrow \Omega_x \sim \Omega_y .$$

<u>Lemma 1.5.5.</u>- Let x be an element of the maximal ideal of a curve $\square$, and $\square'$ a local ring such that $k((x)) \subset \square' \subset \overline{\square}$. Then:

i) $\square'$ has only one maximal ideal.

ii) $\square'$ is a $k((x))$-module of finite type. Particulary, $\square'$ is noetherian.

iii) $\square'$ is the local ring of an irreducible algebroid curve over k.

<u>Proof:</u> i) The ring extension $\overline{\square}/_{k((x))}$ is integral. If $\underline{m}'_1$, $\underline{m}'_2$ are two maximal ideals of $\square'$, there exist two maximal ideals $\overline{m}_1$, $\overline{m}_2$ in $\overline{\square}$ such that $\overline{m}_i \cap \square' = \underline{m}'_i$, $i=1,2$. Since $\overline{\square}$ is local, $\overline{m}_1 = \overline{m}_2$, whence $\underline{m}'_1 = \underline{m}'_2$.

ii) $\overline{\square}$ is a $k((x))$-module of finite type and $\square'$ is a $k((x))$-submodule of $\overline{\square}$, then $\square'$ is itself a $k((x))$-module of finite type.

iii) $\square'$ is a local noetherian domain. Furthermore, since it is integral over $k((x))$, we have

$$\dim \square' = \dim k((x)) = 1 .$$

We must prove that $\square'$ is complete for its Krull topology, and that it has k as a coefficient field.

Let m' and q be the maximal ideals of $\square'$ and $k((x))$ respectively, and denote by q' the extended ideal of q by the homomorphism $k((x)) \longrightarrow \square'$. Since $\operatorname{Spec}(\square') = \{(0), m'\}$, we have $\sqrt{q'} = m'$ and thus q' is m'-primary.

The q-adic topology over $\square'$, considered as $k((x))$-module, is given by the filtration $q\square' \supset q^2\square' \supset \ldots \supset q^n\square' \supset \ldots$; i.e. by the filtration $q' \supset q'^2 \supset \ldots \supset q'^n \supset \ldots$ This is actually its q'-adic topoloy as ring, and since q' is m'-primary and $\square'$ noetherian, it is also its m'-adic topology.

Consider the q-adic topology over the $k((x))$-module $\square'$. $k((x))$ is a complete submodule and $\square'$ is free of finite type over

$k[[x]]$, then $\square'$ is complete.

It only remains for us to see that $\square'$ has k as a coefficient field. The canonical map $k[[x]] \longrightarrow \square'/m'$ induces a homomorphism

$$k = k[[x]]/q \longrightarrow \square'/m'$$

which is actually an isomorphism since $\square'/m'$ is algebraic over k.

Proposition 1.5.6.- Let $\square$ be the local ring of an irreducible algebroid curve and x a transversal parameter. Then $\square_x$ is the only ring of type $\square_y$ ($y \in \underline{m} - \{0\}$) such that the map

$$\psi_y : \mathrm{Spec}(\square_y) \longrightarrow T$$
$$\Omega_y \longmapsto \Omega_y\sim$$

is bijective.

Proof: Since x is transversal we have $\square \subset \square_x \subset \overline{\square}$ (see 1.4.3.). By the above lemma $\mathrm{Spec}(\square_x)$ consists of only two ideals, (0) and the maximal ideal $\underline{m}_x$. Hence, by 1.5.5. T has exactly two elements. If $y \in \underline{m} - \{0\}$ is an element such that ψ_y is bijective, $\mathrm{Spec}(\square_y)$ consists of two ideals, (0) and $\underline{m}_y$. Then $\underline{m}_y$ is maximal and $\square_y$ is local. Since $(\square_y)_{\underline{m}_y} = \square_y$ and $(\square_y)_{(0)} = F$ and ψ_y is bijective, it is clear that $\psi_y((0)) = \psi_x((0))$ and $\psi_y(\underline{m}_y) = \psi_x(\underline{m}_x)$. The last of these equalities asserts $\square_x = (\square_x)_{\underline{m}_x} = (\square_y)_{\underline{m}_y} = \square_y$.

Definition 1.5.7.- Let $\square$ be an irreducible algebroid curve and x a transversal parameter for $\square$. The <u>(strict) quadratic transform of $\square$</u> is defined to be the curve given by the ring $\square_1 = \square_x$.

The above results show that $\square_1$ is an irreducible algebroid curve over the same field, which does not depend on the choice of the transversal parameter. Note, in particular, that the quadratic transform is intrinsecally constructed from the ring $\square$.

Theorem 1.5.8.- Let $\mathcal{O}$ be a curve and $\mathcal{O}_1$ its quadratic transform. The following properties hold:

i) $\mathcal{O}_1$ is the local ring of some irreducible algebroid curve.

ii) $\mathcal{O}_1$ is integral over $\mathcal{O}$, i.e., $\mathcal{O} \subset \mathcal{O}_1 \subset \overline{\mathcal{O}}$. In particular $\overline{\mathcal{O}}_1 = \overline{\mathcal{O}}$ and the associated valuation to $\mathcal{O}$ coincides with the associated valuation to $\mathcal{O}_1$.

iii) Let $B = \{x_i\}_{1 \leqslant i \leqslant N}$ a basis of $\underline{m}$, where x_1 is transversal. There exists a set $\{a_i\}_{1 \leqslant i \leqslant N} \subset k$ such that

$$B_1 = \{ x_1, \frac{x_2}{x_1} - a_2, \ldots, \frac{x_N}{x_1} - a_N \}$$

is a basis of the maximal ideal $\underline{m}_1$ of $\mathcal{O}_1$.

iv) If we have a parametric representation for $\mathcal{O}$ in the basis B given by

$$(1) \qquad x_i = x_i(t) \quad , \quad 1 \leqslant i \leqslant N,$$

then, a parametric representation for $\mathcal{O}_1$ is given by

$$(2) \qquad \begin{array}{l} x_1 = x_1(t) \\ x'_i = x_i(t)/x_1(t) \quad , \quad 2 \leqslant i \leqslant N. \end{array}$$

The centre of representation (2) needs not be $(0,\ldots,0)$.

v) With $\mathcal{O}$, $\mathcal{O}_1$, B, B_1, x_i and x'_i as above, we have

$$\mathcal{O}_1 = k((x_1))\{x'_2,\ldots,x'_N\} \ , \ \mathrm{Emb}(\mathcal{O}_1) \leqslant \mathrm{Emb}(\mathcal{O}) \ , \ e(\mathcal{O}_1) \leqslant e(\mathcal{O}).$$

vi) The contracted ideal of the ideal $\underline{p}_1$ defining $\mathcal{O}_1$, by the homomorphism

$$\begin{array}{rcl} T : k((X_1,\ldots,X_N)) & \longrightarrow & k((X'_1,\ldots,X'_N)) \\ X_1 & \longmapsto & X'_1 \\ X_i & \longmapsto & X'_1 . (X'_i + a_i), \quad i \geqslant 2 \end{array}$$

is the ideal $\underline{p}$ defining $\square$.

Proof: i) If $x \in \underline{m}$ is transversal, it is minimally valuated for $\underline{v}$ in $\underline{m}$, then $\forall z \in \underline{m}$, $\underline{v}(z/x) \geq 0$, whence $z/x \in \overline{\square}$. Thus $\square_1 \subset \overline{\square}$.

ii) $\square \subset \square_1 \subset \overline{\square} \implies \overline{\square}_1 = \overline{\square}$.

iii) Denote by a_i the residue module $\underline{m}_1$ of x_i/x_1, $i \geq 2$. Now,

$$\square_1 = \square\left(\frac{x_2}{x_1}, \ldots, \frac{x_N}{x_1}\right) = \square\left(\frac{x_2}{x_1} - a_2, \ldots, \frac{x_N}{x_1} - a_N\right),$$

and since $x_i = \left(\frac{x_i}{x_1} - a_i\right).x_1 + a_i x_i$, B_1 is a basis of $\underline{m}_1$.

iv) It is trivial from iii).

v) $\square_1 = \square\left(\frac{x_2}{x_1}, \ldots, \frac{x_N}{x_1}\right) = k((x_1))(x'_2, \ldots, x'_N)$. If in iii) the basis B is taken to be minimal, we get $\mathrm{Emb}(\square_1) \leq \mathrm{Emb}(\square)$. Finally, as $e(\square_1)$ is the minimum of values of $\underline{v}$ on $\underline{m}_1$ and $x_1 \in \underline{m}_1$, trivially $e(\square_1) \leq e(\square)$.

vi) Denote by $k((X'_1, \ldots, X'_N)) \xrightarrow{w} k((t))$ the canonic homomorphism defined by the parametrization (2) of iv). Then $w.T$ is the parametrization (1), and hence

$$\underline{p} = (w.T)^{-1}(0) = T^{-1}(w^{-1}(0)) = T^{-1}(\underline{p}_1).$$

This completes the proof of vi) and therefore of the theorem.

Lemma 1.5.9.- For a curve $\square$ the conditions

a) $\square$ is regular,

b) $\square = \square_1$,

are equivalent.

Proof: a) $\Rightarrow$ b) $\square$ is regular is equivalent to $\square = \overline{\square}$. Since $\square \subset \square_1 \subset \overline{\square}$, actually $\square_1 = \square$.

b) $\Rightarrow$ a) Take a transversal parameter x for $\square$. If $\square$ is not regular, $\mathrm{Emb}(\square) > 1$, then the set $\{x\}$ is not a basis of $\underline{m}$, and therefore there exists $y \in \underline{m} - \square x$. Now, $y/x \in \square_1$ and $y/x \notin \square$ then $\square \neq \square_1$.

Theorem 1.5.10.- (RESOLUTION OF SINGULARITIES). Let $\square$ be a curve and consider the ascending chain of successive quadratic transforms starting at $\square$:

$$\square \subset \square_1 \subset \ldots \subset \square_i \subset \square_{i+1} \subset \ldots \subset \overline{\square}.$$

This chain is stationary and it finishes in $\overline{\square}$.

Proof: $\overline{\square}$ is a noetherian $\square$-module, and each $\square_i$ is a $\square$-submodule of $\overline{\square}$. The chain is therefore stationary. If M denotes the first integer for which $\square_M = \square_{M+1}$, the above lemma states that $\square_M$ is regular and thus

$$\square_M = \overline{\square}_M = \overline{\square}.$$

This completes the proof.

Definition 1.5.11.-Two curves defined over the same field are said to be equiresoluble when the sequences formed respectively by the multiplicities of the rings in their desingularization chains 1.5.10. are both identical.

Remark 1.5.12.- The idea of equiresolution, which is based on that of the quadratic transformation, defined in an intrinsic manner in 1.5.7., is essential to classify singularities of curves, as it will be made clear later. In order to best illustrate these aspects, quadratic transformations will be studied now from a geometrical point of view and for embedded curves.

In fact, a quadratic transformation T will be considered

as a certain transformation of N-spaces. Thus, for any embedding of a curve $\square$ in an N-space, the curve $\square_1$ defined in 1.5.7. will be the image of $\square$ by T.

This idea can be formalized by using the language of schemes, because, as we have pointed out in 1.1.3., the notion of "embedding" has a precise meaning there.

Consider the N-space Spec(A) where $A = k[[X_1, \ldots, X_N]]$. Since A is local, Spec(A) has a unique closed point, its maximal ideal M, which sometimes by geometrical reasons will be denoted by O.

The ring $\bigoplus_{n=o}^{\infty} M^n$ is a graded A-algebra of finite type, hence $\mathrm{Proj}(\bigoplus_{n=o}^{\infty} M^n)$ is a finite type projective Spec(A)-scheme which is denoted by $\mathrm{Bl}_M(A)$ (or $\mathrm{Bl}_O(\mathrm{Spec}(A))$). The morphism of k-schemes $\pi : \mathrm{Bl}_M(A) \longrightarrow \mathrm{Spec}(A)$ is called the <u>global blowing up of Spec(A) with centre O.</u>

We need some well known basic properties of $\mathrm{Bl}_M(A)$, (Bennet, [5], or Romo [19]):

1) $\mathrm{Bl}_M(A) = \bigcup_{i=1}^{N} D_+(X_i)$, where $D_+(X_i)$ is an affine open set isomorphic to

$$\mathrm{Spec}\left(A\left[\frac{X_1}{X_i}, \ldots, \frac{X_N}{X_i} \right] \right),$$

$\pi|_{D_+(X_i)}$ is the morphism induced by the inclusion

$$A \longrightarrow A\left[\frac{X_1}{X_i}, \ldots, \frac{X_N}{X_i} \right] ,$$

and the $D_+(X_i)$ glueing together in the obvious manner.

2) $\pi^{-1}(O)$ is a closed k-subscheme of $\mathrm{Bl}_M(A)$ (and actually a divisor) isomorphic to the projective space $\mathrm{Proj}(\bigoplus_{n=o}^{\infty} M^n/M^{n+1}) = \mathrm{Proj}(k[\bar{X}_1, \ldots, \bar{X}_N])$, $\bar{X}_i = X_i + M^2$, and the canonical embedding

$$\begin{array}{ccc} \pi^{-1}(M) & \longrightarrow & Bl_M(A) \\ \wr\wr & & \| \\ \mathrm{Proj}(\bigoplus_{n=o}^{\infty} M^n/M^{n+1}) & \longrightarrow & \mathrm{Proj}(\bigoplus_{n=o}^{\infty} M^n) \end{array}$$

is induced by the natural graded ring homomorphism

$$\bigoplus_{n=o}^{\infty} M^n \longrightarrow \bigoplus_{n=o}^{\infty} M^n/M^{n+1} .$$

Thus, the closed points of $\pi^{-1}(O)$ correspond one-one and onto to the directions through the origin in an affine N-space, and such a point O' may be represented by its homogeneous coordinates $(a_1,\ldots,a_N)$. The scheme $\pi^{-1}(O)$ is called the <u>exceptional divisor of π.</u>

3) A point $O'=(a_1,\ldots,a_N)\in\pi^{-1}(O)$ is in $D_+(X_i)$ if and only if $a_i \neq 0$. Moreover, O' is a regular point of $Bl_M(A)$ and

$$\{\frac{X_1}{X_i}-\frac{a_1}{a_i},\ldots,X_i,\ldots,\frac{X_N}{X_i}-\frac{a_N}{a_i}\}$$

is a regular system of parameters for the local ring $\mathcal{O}_{Bl_M(A),O'}$. Thus, setting $Z_j=\frac{X_j}{X_i}-\frac{a_j}{a_i}$, $j\neq i$, and $Z_i = X_i$, the complection $\hat{\mathcal{O}}_{Bl_M(A),O'}$ of $\mathcal{O}_{Bl_M(A),O'}$ is isomorphic to $k((Z_1,\ldots,Z_N))$ and the homomorphism of local rings

$$T : k((X_1,\ldots,X_N)) \longrightarrow k((Z_1,\ldots,Z_N)),$$

induced by π is given by

$$\begin{array}{lll} X_j \longmapsto & Z_iZ_j + \frac{a_j}{a_i} Z_i , & j\neq i , \\ X_i \longmapsto & Z_i . & \end{array}$$

<u>Definitions 1.5.13.-</u> The homomorphism T (or its analogous scheme theoretical morphism T^*: $\mathrm{Spec}(k((Z_1,\ldots,Z_N)))\longrightarrow \mathrm{Spec}(k((X_1,\ldots,X_N)))$ is called <u>a formal quadratic transformation</u> in the direction $(a_1,\ldots,a_N)$.

Thus, a formal quadratic transformation may be viewed as a transformation of N-spaces.

On the other hand, it follows from the above construction that $Z_i \in \mathcal{O}_{Bl_M(A),O'}$ is a local equation for the exceptional divisor in the open $D_+(X_i)$, and so the <u>exceptional divisor of T</u> may be defined to be the hyperplane $Z_i=0$ (i.e., the subscheme $Spec(k[[Z_1,\dots,Z_N]]/(Z_i))$ of $Spec(k[[Z_1,\dots,Z_N]])$). Intuitively, O is blowed up into $Z_i=0$ by T.

<u>Remarks and definitions 1.5.14.-</u> Let C be an embedded irreducible algebroid curve in the N-space Spec(A), and let

$$x_i = x_i(t) \quad , \quad 1 \leqslant i \leqslant N ,$$

be parametric equations for C. Assume that $X_1=0$ is transversal to C, i.e., x_1 is a transversal parameter for the curve.

From 1.2.4. and 1.2.5. it is easy to see that the closed point $O' \in \pi^{-1}(O)$ determined by the direction of the tangent line to C is $O'=(1,a_2,\dots,a_N)$, where $a_j = x_j(t)/x_1(t) \pmod{(t)}$. Moreover, by 1.5.8. vi) the formal quadratic transformation

$$\begin{array}{llcll} T: & A=k[[X_1,\dots,X_N]] & \longrightarrow & k[[X'_1,\dots,X'_N]] \cong \hat{\mathcal{O}}_{Bl_M(A),O'} & \\ (1) & X_1 & \longmapsto & X'_1 & \\ & X_j & \longmapsto & X'_j X'_1 + a_j X'_1 & , \ j \geqslant 2, \end{array}$$

verifies $T^{-1}(\underline{p}) = \underline{p}_1$, and so it induces a commutative diagram of schemes as follows

$$\begin{array}{lccc} & Spec(\square_1) & \xrightarrow{\ i_1\ } & Spec(k[[X'_1,\dots,X'_N]]) \\ (2) & Q \downarrow & & \downarrow T^* \\ & Spec(\square) & \xrightarrow{\ i\ } & Spec(k[[X_1,\dots,X_N]]) \end{array}$$

where i and i_1 are the embeddings given respectively by $\underline{p}$ and $\underline{p}_1$, and Q is the morphism induced by the inclusion $\square \longrightarrow \square_1$.

Therefore, the embedded curve C given by $\underline{p}$ is transformed by T into the embedded curve C' (in $\mathrm{Spec}(k[[X'_1,\dots,X'_N]])$ given by $\underline{p}_1$, so C' will be called the strict quadratic transform of C by T, (compare this definition with 1.5.7.). The total quadratic transform of C is defined to be the union of C' and the exceptional hyperplane $X'_1=0$.

Note that when $N=2$, the total quadratic transform of C is formed by two curves, C' and $X'_1=0$, i.e., it is a "reducible curve". Reducible curves are not considered in this work, by reasons which are explained in Chapter III,3.1., but they will be mentioned seldomly in order to clarify certain ideas such as that the total transform of a plane curve.

To give an embedded (reducible in general) algebroid curve C in an N-space means to give an ideal $\underline{q} \subset k[[X_1,\dots,X_N]]$ with $\sqrt{\underline{q}} = \underline{q}$ and depth $\underline{q} = 1$. Thus $\underline{q}$ has an irredundant primary decomposition $\underline{q} = \underline{p}_1 \cap \dots \cap \underline{p}_m$ where $\underline{p}_i$ is prime, and since $\dim(k[[X_1,\dots,X_N]]/\underline{q}) = 1$ (see 1.1.7. iv)) it is clear that depth $\underline{p}_i=1$. The embedded irreducible algebroid curves Γ_i defined by $\underline{p}_i$ are called irreducible components of C, and we may write $C = \Gamma_1 \cup \dots \cup \Gamma_m$.

Now, assume $X_1=0$ is transversal to C, i.e., $X_1=0$ is transversal to each Γ_i. Let $\Delta(C) = \{t_1,\dots,t_s\}$ be the set of distinct tangent lines to the curves Γ_i (i.e., the tangent cone of C). Each t_j determines a point $O'(j)$ in the exceptional divisor and a formal quadratic transformation $T(j)$. The union of the strict quadratic transforms Γ'_i by $T(j)$ of all Γ_i with tangent t_j is noted by C'_j. In the same way, we denote by C^*_j the union $C'_j \cup E_j$ where E_j is the exceptional divisor of $T(j)$. Then the strict (rep. total) quadratic transform of C is defined to be

$$T_{st}(C) = \bigcup_{t_j \in \Delta(C)} C'_j \qquad (\text{resp. } T_{tot}(C) = \bigcup_{t_j \in \Delta(C)} C^*_j),$$

and C'_j (resp. C^*_j) are called connected components of $T_{st}(C)$ (resp. $T_{tot}(C)$). (In fact $T_{st}(C)$ and $T_{tot}(C)$ are subsets of a topological disjoint union of N-spaces, and C'_j and C^*_j are respectively the connected components of these subsets.)

Remark 1.5.14.- In 1.5.8. vi) it is given the behaviour of the prime ideal $\underline{p}$ defining an embedded irreducible algebroid curve C under a formal quadratic transformation. Assume that C is plane and hence $\underline{p}$ is principal. Take a generator $f(X,Y)$ of $\underline{p}$ and set

$$f(X,Y) = f_n(X,Y) + f_{n+1}(X,Y) + \dots ,$$

where f_j is a homogeneous polynomial of degree j and $f_n \neq 0$. Since f is irreducible, f_n is a power of a linear form and thus, if $X=0$ is transversal to C,

$$f_n(X,Y) = c\,(Y - aX)^n \quad , \qquad c, a \in k \text{ and } c \neq 0.$$

We consider the point $O' = (1,a) \in \pi^{-1}(O)$ determined by the tangent line $Y - aX = 0$, and the formal quadratic transformation T given by

$$X = X'$$
$$Y = Y'X' + aX'.$$

Then we have $f(X,Y) = X'^n f'(X',Y')$, where

$$f'(X',Y') = Y'^n + X' f_{n+1}(1, Y'+a) + X'^2 f_{n+2}(1, Y'+a) + \dots$$

The series $f'(X',Y')$ is a generator of the ideal defining C' and it is constructed directly from $f(X,Y)$ and T, so this series may be called the <u>strict quadratic transform of f by T</u>.

Notations and definitions 1.5.15.- In 1.5.10. is showed that a sequence of quadratic transformations may be considered in order to desingularize a curve □. Sequences of quadratic transformations are studied classically by using the language of infinitely near points. Sometimes we shall use this language and so the basic definitions for it will be introduced now from a modern view point.

Let O be a closed point of an N-space Spec(A). An <u>infinitely near point of O in its first neighbourhood</u> is defined to be a closed point O_1 in the exceptional divisor of the blowing up of Spec(A) with centre O. For such a O_1, set $A(O_1) = \hat{\mathcal{O}}_{Bl_M(A), O_1}$ and M_1 the maximal ideal of $A(O_1)$. By recurrence, if O_i is an infinitely near point of O in its i-th order neighbourhood, if $A(O_i)$ is defined and if M_i denotes the maximal ideal of $A(O_i)$, we define an infinitely near point of O in its (i+1)-order neighbourhood to be an infinitely near point O_{i+1} of the closed point $\overline{O}_i$ ($=M_i$) of $Spec(A(O_i))$ in its first order neighbourhood, and we set

$$A(O_{i+1}) = \hat{\mathcal{O}}_{Bl_{M_i}(A(O_i)), O_{i+1}}$$

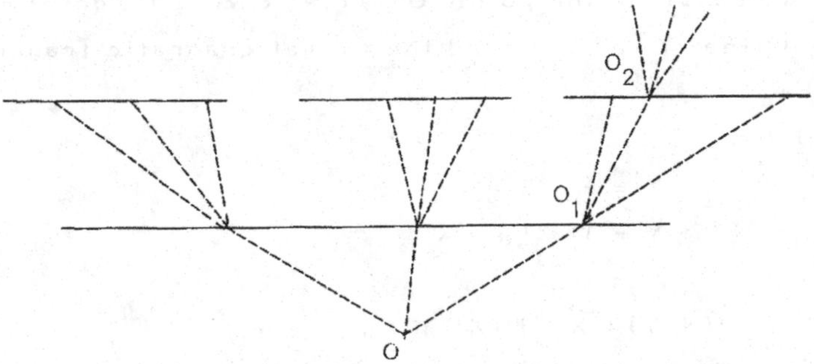

For a sequence of infinitely near points we means a sequence $O_o, O_1, \ldots, O_i, \ldots$ such that O_{i+1} is a point in the first neighbourhood of $\overline{O}_i$ for all $i \geq 0$.

Now, let C be an irreducible algebroid curve in the N-space Spec(A).

<u>Proposition-Definition 1.5.16.</u>- The curve C determines uniquely a sequence $O_o, O_1, \ldots, O_i, \ldots$ of infinitely near points which is called the <u>sequence of infinitely near points of the origin O of C</u> (the origin O is the closed point of Spec(□) identified with the closed point of Spec(A) by the embedding given by C). Conversely this sequence determines completely the embedded curve C.

<u>Proof</u>: Let $O_o = O$. The embedding given by C,

$$\mathrm{Spec}(\square) \xrightarrow{\ i\ } \mathrm{Spec}(A)$$

corresponds to an epimorphism of local rings $A \xrightarrow{i^*} \square$, which induces a homomorphism of graded rings

$$\bigoplus_{n=o}^{\infty} M^n/_{M^{n+1}} \xrightarrow{\ \bar{i}^*\ } \bigoplus_{n=o}^{\infty} \underline{m}^n/_{\underline{m}^{n+1}}$$

where M (resp. $\underline{m}$) is the maximal ideal of A (resp. □). On applying the functor Proj to the sequence

$$\bigoplus_{n=o}^{\infty} M^n \xrightarrow{\ \bar{j}\ } \bigoplus_{n=o}^{\infty} M^n/_{M^{n+1}} \xrightarrow{\ \bar{i}^*\ } \bigoplus_{n=o}^{\infty} \underline{m}^n/_{\underline{m}^{n+1}}$$

($\bar{j}$ is the natural map), we obtain the sequence

$$\begin{array}{ccccc} \mathrm{Proj}\big(\bigoplus_{n=o}^{\infty} \underline{m}^n/_{\underline{m}^{n+1}}\big) & \xrightarrow{\mathrm{Proj}(\bar{j})} & \mathrm{Proj}\big(\bigoplus_{n=o}^{\infty} M^n/_{M^{n+1}}\big) & \xrightarrow{\mathrm{Proj}(\bar{i}^*)} & \mathrm{Proj}\big(\bigoplus_{n=o}^{\infty} M^n\big) \\ & & \wr\!\shortparallel & & \shortparallel \\ & & \pi^{-1}(O) & . & Bl_M(A) \end{array} .$$

$\mathrm{Proj}(\bar{j})$ is the inclusion of $\pi^{-1}(O)$ in $Bl_M(A)$ (1.5.12.) and $\mathrm{Proj}\big(\bigoplus_{n=o}^{\infty} \underline{m}^n/_{\underline{m}^{n+1}}\big)$ has only a point (see 1.2.5.), then its image by $\mathrm{Proj}(\bar{i}^*)$ is a point O_1, infinitely near of O, as desired. Furthermore the diagram (2) in 1.5.14. takes the form

$$(3)\qquad \begin{array}{ccc} \mathrm{Spec}(\square_1) & \xrightarrow{\;i_1\;} & \mathrm{Spec}(A(O_1)) \\ {\scriptstyle Q}\downarrow & & \downarrow{\scriptstyle T^*} \\ \mathrm{Spec}(\square) & \xrightarrow{\;i\;} & \mathrm{Spec}(A) \end{array}$$

and the rest of the sequence is constructed by induction.

Conversely, let $O_o, \ldots, O_i, \ldots$ and $O'_o, \ldots, O'_i, \ldots$ be respectively the sequences for C and C', and assume $O_i = O'_i$ for all i. We claim that C=C'. In fact, by diagram (3) we only need to prove the claim for C and C' regular. In this case, if $A = k[[X_1, \ldots, X_N]]$ and if $X_1 = 0$ is transversal to C, then $X_1 = 0$ is transversal to C', because $O_1 = O'_1$. Therefore, we have parametric equations

$$C : \begin{array}{l} x_1 = x_1 \\ x_j = a_{j1} x_1 + a_{j2} x_1^2 + \ldots \end{array} \qquad C' : \begin{array}{l} x_1 = x_1 \\ x_j = a'_{j1} x_1 + a'_{j2} x_1^2 + \ldots \end{array}$$

$$1 \leq j \leq N.$$

Now, since $O_i = O'_i$, one may prove easily that $a_{ji} = a'_{ji}$ for all i, j, and hence C=C'.

<u>Definition 1.5.17.</u>- The <u>multiplicity $e(O_i)$</u> of the point O_i in the sequence of infinitely near points of the origin of an irreducible curve is defined to be the multiplicity of the ring $\square_i$, i-th quadratic transform of $\square$.

Sometimes, for simplicity, we shall say that the curve <u>lies</u> on the points $O_o, O_1, \ldots, O_i, \ldots$, and that it has multiplicity $e(O_i)$ at O_i.

<u>Remark 1.5.18.</u>- For a reducible curve C, taking its irreducible components and the sequences of infinitely near points determined by them, one obtains a <u>tree</u> of infinitely near points, uniquely constructed from C. To each point of this tree one may attach a multiplicity, by adding the multiplicities of the irreducible components in it. Since the components are different, by 1.5.16., these multiplicities became 1 for the order of the neighbourhood large enough.

If O_i is a point of the tree in the i-th order neighbourhood, then it follows from 1.5.16. that for each component Γ_l lying on O_i, the i-th strict quadratic transform of the local ring of Γ_l is actually embedded in the N-space $Spec(A(O_i))$. Thus the <u>i-th strict quadratic transform of C may be defined intrinsically</u>, without making reference to formal quadratic transformations: its connected component at O_i is the embedded curve in $Spec(A(O_i))$ whose irreducible components are the i-th transform of the Γ_l lying on O_i.

Moreover, by the observation made above, for i large enough, O_i is a regular point for the i-th strict transform. (If C is plane, the i-th total transform may be defined in this way, and for i large enough, O_i is an ordinary double point, i.e., a normal crossing of two regular curves).

To end this chapter we shall introduce satellite and proximate points for the case N=2 which will be often used in the sequel.

Let O be the closed point of a 2-space (=plane). Take an infinitely near point O_i of O in its i-th order neighbourhood. Let $E(O_i)$ the total exceptional curve at O_i, i.e, $E(O_1)$ is the exceptional divisor of π at O_1, and $E(O_i)$ is the connected component at O_i of the (i-1)-th total quadratic transform of $E(O_1)$, for $i \geqslant 2$. It is evident that O_i is a regular or ordinary double point of $E(O_i)$.

<u>Definition 1.5.19.</u>- We say that the infinitely near point O_i of O is <u>free respect O</u> if O_i is a non singular point of $E(O_i)$; and we say that O_i is <u>a satellite of O</u> if it is a singular point (and actually ordinary double) of $E(O_i)$.

<u>Definition 1.5.20.</u>- The satellite point O_i is called a <u>terminal satellite of O</u> if O_{i+1} is free respect O. The point O_{i+1} is then called a <u>leading free point of O</u>.

<u>Definition 1.5.21.</u>- Given a point O_i in the sequence of infinitely

near points of the origin of an embedded irreducible plane curve C, we say that $O_{i+1}, O_{i+2}, \ldots, O_{i+r}$ are <u>proximate points of O_i</u> when

$$e(O_i) = e(O_{i+1}) + e(O_{i+2}) + \ldots + e(O_{i+r}) .$$

CHAPTER II

HAMBURGER-NOETHER EXPANSIONS OF ALGEBROID CURVES.

The purpose of this chapter is to give the technical material which will be used in the following chapters. In the first section we shall see that there is no possibility of obtaining Puiseux expansions in positive characteristic. The Hamburger-Noether expansions, which will replace the Puiseux ones, are developed in the remaining sections.

1. PUISEUX EXPANSIONS. DISCUSSION.

It is well known that every algebroid curve over an algebraically closed field of characteristic zero may be exhibited by a Puiseux series. We shall see how in positive characteristic this is not at all true, and the classical methods are not sufficient.

Let us consider a primitive parametric representation of an irreducible algebroid curve $\square$ over an algebracally closed field k:

$$x_i = x_i(t) \quad , \quad \underline{v}(x_i) > 0 \, , \quad 1 \leqslant i \leqslant N.$$

Assume that $N = \mathrm{Emb}(\square)$.

The procedure to get all the primitive parametric representations of $\square$ is as follows. First, if we keep the basis

$\{x_i\}_{1 \leq i \leq N}$ of the maximal ideal $\underline{m}$ fixed, any other primitive parametric representation is obtained by changing the uniformizing parameter by a substitution of type:

$$(1) \qquad t = c_1 t' + c_2 t'^2 + \dots \quad , c_i \in k \, , c_1 \neq 0 .$$

Now, we may fix t, and changing the basis $\{x_i\}_{1 \leq i \leq N}$ by another one $\{x'_j\}_{1 \leq j \leq M}$ $(N \leq M)$ the relation takes the form

$$(2) \qquad x'_j = f_j(x_1, \dots, x_N) \quad , \quad 1 \leq j \leq M$$

where f_j is a series in N variables with order > 0, and the Jacobian matrix of the forms of degree 1 of the series f_i has the maximum rank .(This rank is actually N.)

<u>Remark 2.1.1.</u>- In characteristic zero, for any series $x = x(t)$ in $k((t))$ of order $n > 0$, there is always a substitution of type (1) for which $x = t'^n$ (it is sufficient to use indeterminate coefficients).

Consequently, for each curve a Puiseux type of primitive parametric representation can be obtained in any basis of $\underline{m}$:

$$\begin{aligned} x_1 &= t'^n \\ x_i &= x_i(t') \quad , \quad 2 \leq i \leq N. \end{aligned}$$

If charact. $k = p > 0$ and if $x = x(t) \in k((t))$ is a series of order n, then we can assume the last property only when $(n,p) = 1$. In fact, for instance, for $x = t^p + t^{p+1}$, there is no substitution (1) for which $x = t'^p$.

<u>Example 2.1.2.</u>- <u>A priori</u>, nevertheless, every curve could have a privileged basis of its maximal ideal having a Puiseux expansion relative to a transversal parameter. We give below an example to show that this question has a negative answer.

Let us consider the curve

$$x = t^p + t^{p+1}$$
$$y = t^m \qquad , \qquad m > p+1 \quad ,$$

over a field of characteristic $p > 0$. If x' is a transversal parameter, then:

$$x' = ax + by + f(x,y) \quad , \text{ with } a \neq 0 \text{ and } \underline{\upsilon}(f) \geq 2, \text{ i.e.}$$
$$x' = at^p + at^{p+1} + g(t) \quad , \quad \text{with } \underline{\upsilon}(g) > p+1 \quad , \quad a \neq 0 .$$

Hence we conclude that it is not possible to write $x' = t'^p$ for any uniformizing parameter t'.

Moreover, the curve

$$x = t^{p^2} + t^{p^2+1}$$
$$y = t^{p^2+p} + t^{p^2+p+1}$$

is such that no element in $\underline{m} - \underline{m}^2$ can be expressed as t'^n for any uniformizing t' and for any integer $n > 0$.

Any extension of type $k((t))/_{k((x))}$, $x = x(t)$, $\underline{\upsilon}(x) \geq 1$, is separable if and only if $(dx/dt) \neq 0$ (proposition 1.3.14.). In characteristic zero it is always a Galois extension and so it is cyclic. In positive characteristic this is not true. We shall give some examples in the sequel.

Remark 2.1.3.- Let us consider a curve $\square$ and a basis $\{x_i\}_{1 \leq i \leq N}$ of $\underline{m}$ where x_1 is transversal. There always exists some x_i , $1 < i \leq N$, which is separable, because the representation in function of a uniformizing parameter is primitive. Therefore $x_1 + \alpha x_i$ is transversal and separable, except for a finite set of values of α in k.

Since k is infinite we can take a basis containing separable and transversal parameters. Unfortunately these bases are not stable under quadratic transformations, as it is made evident in the following:

<u>Example 2.1.4.</u>- Let k be an algebraically closed field of characteristic $p > 0$ and let □ be the curve over k given by

$$x = t^{p^2} + t^{p^2+1}$$
$$y = t^{p^2+p} + t^{p^2+p+1} .$$

The element x in the basis $\{x, y\}$ of $\underline{m}$ is transversal and separable but the unique transversal element in the basis $\{x, y/x\}$ of the maximal ideal of the quadratic transform of □ is y/x , and it is not separable.

<u>Example 2.1.5.</u>- An extension $k((t))/_{k((x))}$ may not be normal. In fact, let k be a field of characteristic 3 and

$$x = t^3 + t^4 + a_6 t^6 + a_7 t^7 + \ldots$$

We claim that $k((t))/_{k((x))}$ is not normal. Indeed, this is trivial since the identity is the unique $k((x))$-automorphism of $k((t))$.

<u>Remark 2.1.6.</u>- If k again has characteristic 3, for the curve

$$x = t^3 + t^4$$
$$y = t^9 ,$$

any transversal parameter x' is like in the preceding example; thus $k((t))/_{k((x'))}$ is not normal.

Moreover, if we consider the curve

$$x = t^9 + t^{10}$$
$$y = t^{12} ,$$

then, the property "for any $y \in \underline{m} - \underline{m}^2$ $k((t))/_{k((y))}$ is not normal" holds for it.

Remark 2.1.7.- Extensions $k((t))/_{k((x))}$ which are not normal are shown in the above examples. Those examples allow Galoisian extensions whose Galois group is not commutative.

Let $k((t))/_{k((x))}$ be a separable and not normal extension. Let $k((t'))$ be a minimal normal extension of $k((x))$ containing $k((t))$. Trivially $k((t'))/_{k((x))}$ is separable and thus Galoisian. But it cannot be Abelian; otherwise the subextension $k((t))/_{k((x))}$ would be normal which is not true by hypothesis.

Now, if $x = x(t')$ is the expansion of x as a series in t', we may obtain a primitive parametric representation for a curve in which x appears:

$$x = x(t')$$
$$y = t'\, x(t').$$

For a more explicit example, consider the curve

$$x = t^6 + t^8$$
$$y = t^9$$

in characteristic 3, which has the peculiarity that for any transversal parameter x' the extension $k((t))/_{k((x'))}$ is Galoisian but it is not Abelian.

Finally, notice that the assumption that k has characteristic 3 in the above examples is not essential. If charact.$k = p > 0$, by replacing $t^3 + t^4$ by terms of type $t^{p^i} + t^{p^{i+1}}$ one may obtain identical ones.

Now, we try to give a characterization of curves for which Puiseux series exist. Also we shall give requirements in order to know when a Puiseux representation is primitive. In the sequel we shall assume that charact.$k = p > 0$.

Proposition 2.1.8.- Let $x \in k((t))$ with $\underline{\upsilon}(x) = n > 0$. Then, there exists $t' \in k((t))$ satisfying $x = t'^n$ if and only if $(n_s(x),p)$ is 1 (see 1.3.13.).

Proof: Necessity is obvious from 1.3.14. Conversely, let $n'=n_s(x)$ and $k_s = k((\tau))$ be the greatest separable extension of $k((x))$ which is contained in $k((t))$. Then we can choose τ such that $\tau^{n'} = x$, because $(n',p) = 1$.

On the other hand, the inseparability index of τ is $i = i(x)$ and so $\tau = t'^{p^i}$ for some $t' \in k((t))$. Then $x = t'^n$ and thus the proof is complete.

Corollary 2.1.9.- Let $\{x_i\}_{1\leq i \leq N}$ be a basis of the maximal ideal of a curve □. Then □ has a Puiseux expansion (relative to x_1):

$$\begin{aligned} x_1 &= t'^n \\ x_i &= x_i(t') \quad , \quad 2 \leq i \leq N, \end{aligned}$$

in that basis if and only if $(n_s(x_1),p) = 1$.

Corollary 2.1.10.- In order for a curve of multiplicity n to have a Puiseux expansion relative to a transversal parameter in every basis, it is necessary and sufficient that $(n,p)=1$.

Proof: From remark 2.1.3. there always exists a transversal and separable parameter x. Since $n_s(x) = n$, the result follows from the above corollary.

Remark 2.1.11.- Given an equation of a plane curve in a basis $\{x,y\}$ by an irreducible distinguished polynomial over $k((x))$:

$$Y^n + A_{n-1}(X)\,Y^{n-1} + \dots + A_o(X) = 0 \quad , \quad A_i(X) \in k((X)) \; , \quad 0 \leqslant i \leqslant n-1,$$

then the integers $n_s(x)$ and $i(x)$ are respectively the separablity degree and the inseparability index of that polynomial. Consequently, if one knows the equation of a plane curve, this equation determines whether the curve has a Puiseux expansion or not.

A similar remark may be made for twisted curves.

Remark 2.1.12.- The bases in which we have Puiseux expansions are not stable under quadratic transformations. For instance, the curve $\square$ given by

$$x = t^{p^2+1}$$
$$y = t^{p^2+p+1} + t^{p^2+p+2}$$

has a Puiseux expansion relative to any transversal parameter in all the bases of $\underline{m}$ (2.1.10.), but its quadratic transform $\square_1$, whose parametric equations are

$$x = t^{p^2+1}$$
$$y = t^p + t^{p+1}$$

has no Puiseux expansion relative to a transversal parameter in any basis of its maximal ideal.

Proposition 2.1.13.- Let us consider a parametric representation of type

$$(1) \qquad \begin{aligned} x_1 &= t^n \\ x_i &= a_{r_{i1}} t^{r_{i1}} + a_{r_{i2}} t^{r_{i2}} + \dots \; , \quad 2 \leqslant i \leqslant N, \end{aligned}$$

with $1 \leqslant r_{i1} < r_{i2} < \dots$ and $a_{r_{ij}} \neq 0$, $1 \leqslant j < \infty$.

Then the representation is primitive if and only if the g.c.d. of the set

$$(2) \qquad \{n, r_{21}, \ldots, r_{31}, \ldots, r_{N1}, \ldots\}$$

is one. (Note that it needs not have $n \leq r_{i1}$, $2 \leq i \leq N$).

Proof: Necessity is obvious. We shall prove the sufficient condition. If the representation is not primitive, there exists $t' \in k((t))$ with $q = \underline{v}(t') > 1$ such that $x_i \in k((t'))$, $1 \leq i \leq N$. By 2.1.8. the separability degree of the extension $k((t))/_{k((x_1))}$ is prime with the characteristic p of k. It follows that the separability degree of $k((t'))/_{k((x_1))}$ is also prime with p. By using again 2.1.8. we may find t' satisfying $x_1 = t'^{n/q}$.

Now, there is a relation between t and t' of type $t' = w.t^q$, where w is a q-th root of unity in k. In the expansion of x_i as series in t' we may replace t' by $w.t^q$ and get (1). Thus $q > 1$ is a divisor of each element in the set (2) which is contradictory. This completes the proof.

Remark 2.1.14.- One of the fundamental points in the classical study of algebroid curves is the fact that when k is an algebraically closed field of characteristic zero and x is an indeterminate over k, the algebraic closure of $k((x))$ is the union of the fields $k((x^{1/n}))$, $n = 1,2,\ldots$ Hence, every algebraic element over $k((x))$ may be expressed as a formal power series in $x^{1/n}$ with a finite number of negative exponents for some $n > 0$ (see Walker, [22]).

This result is in fact a well known Ostrowski Theorem (Van der Waerden, [20], pag. 53). This theorem was generalized in the case of any characteristic by Ancochea, [2], who proved that every polynomial with coefficients in $k((x))$ factorizes lineary over $k((t))$, by using a linear change like

$$x = t^m + c_1 t^{m+1} + \ldots + c_s t^s \quad , \quad m > 0.$$

Indeed, this generalization is our proposition 1.3.1. (see remark 1.3.7.).

But, as we have seen in this section, in positive characteristic

$$K' = \bigcup_{n=1}^{\infty} k((x^{1/n}))$$

is not algebraically closed. Hence we cannot use Puiseux expansions. G.Rodeja, (12), has proved how in this case Ostrowski's theorem is equivalent to the availability of Hamburger-Noether expansions.

2. HAMBURGER – NOETHER EXPANSIONS OF PLANE ALGEBROID CURVES.

In the preceding section we just saw how Puiseux series are not appropriate in order to study the singularities of algebroid curves in positive characteristic. We shall develop in the sequel Hamburger- Noether expansions, which are the thecnical base in the present work.

For the sake of simplicity we shall begin with the case of plane curves. Recall that k will still denote an algebraically closed field of any characteristic.

Definition 2.2.1.- A parametrization system is defined to be a set $\{x,y\}$ of two formal power series

$$x = x(u)$$
$$y = y(u)$$

in $k((u))$, such that $\underline{v}(x) > 0$ and $\underline{v}(y) > 0$.

Algorithm 2.2.2.- Let us consider a parametrization system $\{x,y\}$ and suppose $n = \underline{v}(x) \leq \underline{v}(y)$. We shall construct the Hamburger-Noether expansion of $\{x,y\}$ by means of an algorithm which is carried out by using successive divisions in $k((u))$ as follows:

Divide y by x and pick out the independent term from the quotient, obtaining a series y_1 of positive order. Continue the division using y_1 instead of y (x always being the denominator) provided that y_1 be a series of order $\geq n$. Stop in the other case. Then one and only one of the two following statements holds:

(A) There exist $h > 0$ and $a_{0i} \in k$, $1 \leq i \leq h$, such that

$$y = a_{01}x + a_{02}x^2 + \ldots + a_{0h}x^h + x^h z_1 , \quad \text{with } 1 \leq \underline{v}(z_1) < \underline{v}(x).$$

(B) There exist $a_{0i} \in k$, $1 \leq i < \infty$, such that

$$y = a_{01}x + a_{02}x^2 + \ldots$$

In the case (A), we continue with the system $\{z_1, x\}$, and again either (A) or (B) is obtained. The process continues as long as (A) is obtained.

Note that when $n = 1$ (B) holds. Hence, after a finite number of steps we are in (B), because the orders of the series z_j are decreasing. This means that an expression as follows is obtained:

$$\text{(D)} \quad \begin{aligned} y &= a_{01}x + a_{02}x^2 + \ldots + a_{0h}x^h + x^h z_1 \\ x &= a_{12}z_1^2 + \ldots + a_{1h_1}z_1^{h_1} + z_1^{h_1}z_2 \\ &\ldots\ldots\ldots\ldots\ldots\ldots\ldots\ldots\ldots\ldots \\ &\ldots\ldots\ldots\ldots\ldots\ldots\ldots \\ z_{r-1} &= a_{r2}z_r^2 + \ldots . \end{aligned}$$

where

$$a_{ji} \in k\ ,\ z_j \in k((u)),\quad 1 \leqslant \underline{v}(z_r) < \ \ldots < \underline{v}(z_1) < \underline{v}(x) \leqslant \underline{v}(y).$$

($a_{j1} = 0$ since $\underline{v}(z_{j-1}) > \underline{v}(z_j)$, $1 \leqslant j \leqslant r$).

<u>Definition 2.2.3.</u>- A Hamburger-Noether expansion for the system $\{x,y\}$ is a set of expresions of type (D) which are verified by it.

<u>Remark 2.2.4.</u>- A Hamburger-Noether expansion provides a parametric representation of a curve, with parameter z_r:

$$x = x(z_r)$$
$$y = y(z_r)$$

Sometimes, we shall denote simply a Hamburger-Noether expansion by an expression like

$$(\underline{D}) \qquad z_{j-1} = \sum_i a_{ji}\, z_j^{\,i} + z_j^{\,h_j} z_{j+1}$$

together with the necessary restrictions on the indices j or the coefficients a_{ji}. For instance (D) will be written as ($\underline{D}$) together with $0 \leqslant j \leqslant r$, $z_o = x$, $z_{-1} = y$ (or simply $0 \leqslant j \leqslant r$ if there is no confusion).

<u>Proposition 2.2.5.</u>- The Hamburger-Noether expansion for a system $\{x,y\}$ with $\underline{v}(x) \leqslant \underline{v}(y)$ is uniquely determined by the conditions:

1) $a_{ji} \in k$.
2) $z_j \in k((u))$.
3) $1 \leqslant \underline{v}(z_r) < \ldots < \underline{v}(z_1) < \underline{v}(x)$.

Particulary, the expansion depends only on the system (and the fixed element in it, if $\underline{v}(x) = \underline{v}(y)$), but it does not depend on the uniformizing u.

<u>Proof</u>: For each expansion of type (D) let $L(D) = h + h_1 + \ldots + h_{r-1}$. For every system $\{x,y\}$ we set

$$L(\{x,y\}) = \min\{L(D) \mid D \text{ is an expansion for } \{x,y\}\}.$$

We shall prove the proposition using induction on $L(\{x,y\})$. If $L(\{x,y\}) = 0$, then there is some expansion (D) for $\{x,y\}$ of type

$$y = a_{01}x + a_{02}x^2 + \ldots .$$

For another expansion (D') one of the following statements would be true

(A) $y = a'_{01}x + a'_{02}x^2 + \ldots + a'_{0h}x^h + x^h z_1$, $1 \leqslant \underline{v}(z_1) < \underline{v}(x)$.

(B) $y = a'_{01}x + a'_{02}x^2 + \ldots$

In the case (A), the equality

$$\sum_{i=1}^{h} (a_{0i} - a'_{0i})\, x^i = \Big(\sum_{i>h} a_i x^i \Big) - x^h z_1$$

becomes a contradiction, because both sides in it are series in u which have different orders.

In the case (B), if q denotes the first integer for which $a_{0q} \neq a'_{0q}$, then again the equality

$$\sum_{i \geqslant q} (a_{0i} - a'_{0i})\, x^i = 0$$

turns out to be a contradiction.

Now, let $\{x,y\}$ be a system such that $L(\{x,y\}) = N > 0$, and suppose that the assertion is true for every system with $L(\{x',y'\}) < N$. Let us consider an expansion (D) for $\{x,y\}$ with $L(D) = N$ and let D' be another expansion. Then $a_{01} = a'_{01}$, since they must agree with the independent term in the series y/x. Hence

$$y_1 = (y - a_{01}x)/_x = (y - a'_{01}x)/_x = y'_1 \; .$$

The system $\{x, y_1\}$ has actually two expansions (D_1) and (D'_1) which are respectively obtained from (D) and (D') by removing the first term. As $L(D_1) = L(D) - 1 = N - 1$, by the induction hypothesis we have $(D_1) = (D'_1)$ and consequently $(D) = (D')$.

<u>Corollary 2.2.6.</u>- The system $\{x, y\}$ is a primitive parametric representation of some curve if and only if for its Hamburger-Noether expansion we have $\underline{\upsilon}(z_r) = 1$.

<u>Proof</u>: If the representation is primitive, the conditions $x(u) \in k((z_r))$ and $y(u) \in k((z_r))$ imply $k((u)) = k((z_r))$, and thus $\underline{\upsilon}(z_r) = 1$. Conversely, if $\underline{\upsilon}(z_r) = 1$ then $k((u)) = k((z_r))$, and since z_r belongs to the quotient field of the curve $\square = k((x(u), y(u)))$ the representation is actually primitive.

<u>Remark 2.2.7.</u>- The preceding corollary is a practical criterion in order to determine when a given parametric representation is primitive.

<u>Definition 2.2.8.</u>- Let $\square$ be an irreducible algebroid plane curve over k and $\{x, y\}$ a basis of the maximal ideal $\underline{m}$ of $\square$. A <u>Hamburger-Noether expansion for $\square$ in this basis</u> is defined to be a Hamburger-Noether expansion of the parametric system $\{x, y\}$.

When a transversal parameter, say x, in that basis is chosen, the Hamburger-Noether expansion in it is completely determined. If there is no confusion, we shall call it the Hamburger-Noether expansion of $\square$ in the basis $\{x, y\}$. Now, we shall find elements in it which depend only on the local ring $\square$. They will make evident the sequence of multiplicities for the singularity of a curve; so that they will give an essential tool to classify singularities.

<u>Proposition 2.2.9.</u>- Let $(\underline{D})$, $0 \leq j \leq r$, $z_0 = x$, $z_{-1} = y$, be a Hamburger-

Noether expansion for the curve $\square$ in the basis $\{x,y\}$. Let $\square_1$ be the quadratic transform of $\square$. Then a Hamburger-Noether expansion for $\square_1$ in the transformed basis $\{x,y_1\}$ ($y_1 = (y-a_{01}x)/x$) is given by

i) If $h > 1$,

$$y_1 = a_{02}x + \dots + a_{0h}x^{h-1} + x^{h-1}z_1$$

$$z_{j-1} = \sum_i a_{ji}z_j^i + z_j^{h_j}z_{j+1} , \quad 1 \leqslant j \leqslant r.$$

ii) If $h = 1$,

$$z_{j-1} = \sum_i a_{ji}z_j^i + z_j^{h_j}z_{j+1} , \quad 1 \leqslant j \leqslant r.$$

The proof proceeds from proposition 2.2.5. by using induction on the minimum number of quadratic transformations which are needed to desingularize $\square$.

Proposition 2.2.10.- Let

$$z_{j-1} = \sum_i a_{ji}z_j^i + z_j^{h_j}z_{j+1} , \quad 0 \leqslant j \leqslant r,$$

the Hamburger-Noether expansion for the curve $\square$ in the basis $\{x,y\}$, $x = z_0$, $y = z_{-1}$. If $\underline{v}$ is the valuation associated to $\square$, and $n_j = \underline{v}(z_j)$, $0 \leqslant j \leqslant r$. Then:

i) The different multiplicites which occur in the desingularization sequence 1.5.10. for $\square$ are exactly the integers n_j , $0 \leqslant j \leqslant r$.

ii) Each n_j is repeated in that sequence exactly h_j times, $0 \leqslant j \leqslant r$, $h_0 = h$, $h_r = \infty$.

iii) The integers h_j and n_j depend only on $\square$.

Proof: iii) is a consequence from i) and ii). We shall prove i) and ii) simultaneously by indution on the minimum number of quadratic transformations which are needed to desingularize $\square$.

For $M(\square) = 0$, the curve is regular and its expansion of type

$$y = a_{01}x + a_{02}x^2 + \dots$$

The result is evident.

Let $\square$ be a curve such that $M = M(\square) > 0$ and let us assume that the result holds for $M(\square') < M$. By applying the induction hypothesis to $\square_1$ and using proposition 2.2.10. it also holds for $\square$.

<u>Corollary 2.2.11.</u>- Let $\square$ and $\square^*$ be curves whose respective Hamburger-Noether expansions are

$$z_{j-1} = \sum_i a_{ji} z_j^i + z_j^{h_j} z_{j+1}, \quad 0 \leqslant j \leqslant r,$$

$$z^*_{j-1} = \sum_i a^*_{ji} z_j^{*i} + z_j^{*h^*_j} z^*_{j+1}, \quad 0 \leqslant j \leqslant r^*.$$

Then, $\square$ and $\square^*$ are equiresoluble iff $r=r^*$, $h_j=h^*_j$, and $n_j=n^*_j$, $0 \leqslant j \leqslant r$.

<u>Remark 2.2.12.</u>- It follows from the proofs of 2.2.9. and 2.2.10. that a Hamburger-Noether expansion determines uniquely a sequence of formal quadratic transformations which desingularize the embedded curve C defined by the choice of the basis $\{x,y\}$. Hence if we take an equation $f(X,Y)=0$ for C, then, by 1.5.14., f and (D) determine an equation $f_i(X_i,Y_i)=0$ for the i-th strict quadratic transform of C. Moreover, according 2.2.10., the sequence of infinitely near points of the origin of C is formed by h points of multiplicity n, h_1 points of multiplicity n_1, etc...

<u>Proposition 2.2.13.</u>- Any expansion (D) of Hamburger-Noether type defines a curve $\square$.

<u>Proof</u>: Consider the expansion

$$(\underline{D}) \qquad z_{j-1} = \sum_i a_{ji} z_j^{\,i} + z_j^{\,h_j} z_{j+1}\,, \quad 0 \leqslant j \leqslant r,\ x = z_0, y = z_{-1}.$$

Let $\underline{v}$ be the associated valuation to $k((z_r))$. Since each z_j belongs to that ring, we have parametric equations

$$x = x(z_r)$$
$$y = y(z_r)$$

defining a curve, whose Hamburger-Noether expansion in the basis $\{x, y\}$ is actually ($\underline{D}$).

3. INTERSECTION MULTIPLICITY OF PLANE CURVES.

Let us consider the embedded plane curve C defined by the irreducible series $f(X,Y) \in k((X,Y))$. Let $\square = k((X,Y))/_{(f)}$ be its local ring and let $\overline{\square} = k((t))$ the integral closure of $\square$ in its quotient field. We have a parametric representation

$$x = x(t)$$
$$y = y(t)$$

of the curve.

Now, let D be another embedded curve, not irreducible in general, defined by the series $g(X,Y) \in k((X,Y))$.

Definition 2.3.1.- The intersection multiplicity of C and D is defined to be the number.

$$(C, D) = \underline{v}(g(x(t), y(t))),$$

where $\underline{v}$ is the associated valuation to $\overline{\square}$.

Remark 2.3.2.- The following properties of the intersection multiplicity are trivial:

i) $(C,D) = \infty \iff C$ is an irreducible component of D.

ii) If $D = D_1 \ldots D_m$ with D_i irreducible for each i, then

$$(C,D) = \sum_{i=1}^{m} (C,D_i).$$

iii) If D is irreducible then $(C,D) = (D,C)$.

iv) $(C,D) \geq e(C).e(D)$. The equality holds if and only if C and D have not a tangent in common.

v) If C_1 (resp. D_1) is the strict quadratic transform of C (resp. D) by a quadratic transformation in the ambient plane, then

$$(C,D) = e(C).e(D) + (C_1, D_1).$$

Now, assume that D is irreducible and the Hamburger-Noether expansion for C and D in the respective basis $\{X+(f), Y+(f)\}$ and $\{X+(g), Y+(g)\}$ of their maximal ideals are given by

$$z_{j-1} = \sum_i a_{ji} z_j^i + z_j^{h_j} z_{j+1}, \quad 0 \leq j \leq r,$$

$$z'_{j-1} = \sum_i a'_{ji} z'^i_j + z'^{h'_j}_j z'_{j+1}, \quad 0 \leq j \leq r'.$$

Let s be the greatest integer for which $h = h'$, $h_1 = h'_1, \ldots, h_{s-1} = h'_{s-1}$, and $a_{jk} = a'_{jk}$ for $j < s$ and $k \leq h_j$. Let i be the least index such that $a_{si} \neq a'_{si}$ $(i \leq h_s + 1,\ i \leq h'_s + 1)$. Finally set

$$S = \sum_{j=0}^{s} h_j . n_j . n'_j .$$

The integer $N = h + h_1 + \ldots + h_{s-1} + i - 1$ is the number of infinitely near points that both curves have in common, (see 1.5.16.).

<u>Proposition 2.3.3.</u>- Keeping the notations as above, we have:

i) If $i \leq h_s$ and $i \leq h'_s$,

$$(C,D) = S + i . n_s . n'_s$$

ii) If $i = h'_s + 1$,

$$(C,D) = S + h_s . n_s . n'_s + n'_{s+1} . n_s$$

iii) If $i = h_s + 1$,

$$(C,D) = S + h'_s . n'_s . n_s + n_{s+1} . n'_s$$

<u>Proof</u>: The proof proceeds trivially from property v) in remark 2.3.2., using induction on N.

<u>Remark 2.3.4.</u>- By adding the products of the respective multiplicities of the infinitely near points that two given curves have in common, one gets their intersection multiplicity.

4. HAMBURGER-NOETHER EXPANSIONS FOR TWISTED CURVES.

We shall generalize in this section the Hamburger-Noether expansions for twisted curves. We shall find results analogous to those of the case of plane curves.

Let k be an algebraically closed field and u an indeterminate over k.

<u>Definition 2.4.1.</u>- An <u>N-dimensional parametric system</u> is defined to be a set $\{x_i\}_{1 \leq i \leq N}$ of formal power series in k((u)):

$$x_i = x_i(u) \quad , \quad 1 \leq i \leq N ,$$

such that $\underline{v}(x_i) > 0$, $1 \leq i \leq N$.

Notations 2.4.2.- Let A be a ring and N a positive integer. We denote by $M_N(A)$ (or simply $M(A)$ if there is no confusion) the A-module of $(N-1)\times 1$-matrices with coefficients in A.

If $A = k((u))$, there is a map $\underline{V}$ defined as follows:

$$\underline{V} : M(k((u))) \longrightarrow Z$$

$$Y = \begin{pmatrix} y_2 \\ \vdots \\ y_N \end{pmatrix} \longmapsto \min\{v(y_2),\ldots,v(y_N)\}$$

where $\underline{v}$ is the natural valuation associated with $k((u))$.

Lemma 2.4.3.- The map $\underline{V}$ has the following properties:

i) $\underline{V}(Y) = \infty \iff Y = 0$, $Y \in M(k((u)))$.

ii) $\underline{V}(z.Y) = \underline{v}(z) + \underline{V}(Y)$, $z \in k((u))$, $Y \in M(k((u)))$.

iii) $\underline{V}(Y_1+Y_2) \geq \min(\underline{V}(Y_1),\underline{V}(Y_2))$, $Y_1, Y_2 \in M(k((u)))$.

Algorithm 2.4.4.- Let us consider a parametrization system $\{x_i\}_{1 \leq i \leq N}$ and assume that $n = \underline{v}(x_1) \leq \underline{v}(x_i)$, $2 \leq i \leq N$. Let

$$Y = \begin{pmatrix} x_2 \\ \vdots \\ x_N \end{pmatrix} \in M(k((u))).$$

We shall use successive divisions as in the case $N=2$. We start dividing all the x_i by x_1, $2 \leq i \leq N$, and continue the division provided that all the obtained quotients with independent terms removed be a series of order $\geq n$.

One of two following situations will be true:

(A) There exist $h > 0$; $A_{0i} \in M(k)$, $1 \leq i \leq h$; and $Z_1 \in M(k((u)))$ such that

$$Y = A_{01} x_1 + A_{02} x_1^2 + \dots + A_{0h} x_1^h + Z_1 x_1^h ,$$

with $1 \leq \underline{V}(Z_1) < \underline{v}(x_1) \leq \underline{V}(Y)$.

(B) There exist $A_{0i} \in M(k)$, $1 \leq i < \infty$, such that

$$Y = A_{01} x_1 + A_{02} x_1^2 + \dots$$

In the case (A), since $\underline{V}(Z_1) = n_1 < n = \underline{v}(x_1)$, there is an element z_1 in the matrix Z_1 such that $\underline{v}(z_1) = n_1$. Then, by considering the parametrization system $\{z_1, x, z_{13}, \dots, z_{1N}\}$ where $z_{13}, \dots, z_{1N}$ are the remaining elements in Z_1, the above process may be repeated. More precisely, if

$$\overline{Z}_0 = \begin{pmatrix} x \\ z_{13} \\ \vdots \\ z_{1N} \end{pmatrix}$$

one of two situations as above will be again true:

(A) There exist $h_1 > 0$, $A_{1i} \in M(k)$, $1 \leq i \leq h_1$; and $Z_2 \in M(k((u)))$, such that

$$\overline{Z}_0 = A_{11} z_1 + A_{12} z_1^2 + \dots + A_{1h_1} z_1^{h_1} + Z_2 z_1^{h_1} ,$$

with $1 \leq \underline{V}(Z_2) < \underline{V}(Z_1) = \underline{v}(z_1)$.

(B) There exist $A_{1i} \in M(k)$, $1 \leq i < \infty$, such that

$$\overline{Z}_0 = A_{11} z_1 + A_{12} z_1^2 + \dots .$$

The process continues when the situation (A) holds. Note, as in the case $N=2$, that after a finite number of steps (B) will be obtained, and the algorithm finishes.

This algorithm yields a set of expresions:

$$(D) \quad \begin{aligned} Y &= A_{01} x_1 + \dots + A_{0h} x_1^h + Z_1 x_1^h \\ \overline{Z}_0 &= A_{11} z_1 + \dots + A_{1h_1} z_1^{h_1} + Z_2 z_1^{h_1} \\ &\dots\dots\dots\dots\dots\dots \\ &\dots\dots\dots\dots\dots\dots \\ \overline{Z}_{r-1} &= A_{r1} z_r + \dots . \end{aligned}$$

such that

$A_{ji} \in M(k)$; $Z_j, \overline{Z}_j \in M(k[[u]])$; $z_j \in k[[u]]$ and z_j is an element of the matrices Z_j and $\overline{Z}_j$; $\underline{V}(Z_j) = \underline{v}(z_j)$; and $1 \leqslant \underline{v}(z_r) < \dots < \underline{v}(z_1) < \underline{v}(x_1) \leqslant \underline{V}(Y)$.

In short, we shall write:

$$(\underline{D}) \qquad \overline{Z}_{j-1} = \sum_i A_{ji} z_j^i + Z_{j+1} z_j^{h_j} , \qquad 0 \leqslant j \leqslant r .$$

<u>Remark 2.4.5.-</u> The matrices A_{j1}, $1 \leqslant j \leqslant r$, have the element 0 exactly in the place in which z_{j-1} is placed in $\overline{Z}_{j-1}$.

<u>Definition 2.4.6.-</u> We define <u>a Hamburger- Noether expansion for the system $\{x_i\}_{1 \leqslant i \leqslant N}$</u> to be a set of expresions of type (D) which are verified by it.

<u>Remark 2.4.7.-</u> When $N > 2$ the Hamburger-Noether expansion is not uniquely determined by the system $\{x_i\}_{1 \leqslant i \leqslant N}$ and the element x_1 fixed in it. In fact the expansion depends on the arrangement of

the elements in the matrices. Even, at some step, it may exist another element z_i' in Z_i, $z_i' \neq z_i$ such that

$$\underline{V}(Z_i) = \underline{v}(z_i').$$

However, the choice of z_i determines completely the next row in the Hambuger-Noether expansion up to arrangements. Furthermore that row does not depend on the uniformizing u.

Definition 2.4.8.- Let $\square$ be a twisted curve over the algebraically closed field k, and $B = \{x_i\}_{1 \leq i \leq N}$ a basis of its maximal ideal. If (D) is a Hamburger-Noether expansion for the parametrization system $\{x_i\}_{1 \leq i \leq N}$, we shall say that (D) is a Hamburger-Noether expansion for the curve $\square$ in the basis B.

Remark 2.4.9.- The results explained in the sequel works as in the case N = 2.

2.4.9.1.- The system

$$x_i = x_i(u), \quad 1 \leq i \leq N,$$

is a primitive parametric representation of a curve if and only if for any Hamburger-Noether expansion for that system $\underline{\upsilon}(z_r) = 1$.

2.4.9.2.- Let

$$(\underline{D}) \qquad \overline{Z}_{j-1} = \sum_i A_{ji} z_j^{\,i} + Z_{j+1} z_j^{h_j}, \quad 0 \leq j \leq r,$$

a Hamburger-Noether expansion for the curve $\square$ in the basis $\{x_i\}_{1 \leq i \leq N}$ of its maximal ideal, where $x_1 = z_0$ is a transversal parameter. Then a Hamburger-Noether expansion for the quadratic transform $\square_1$ of $\square$ in the transformed basis (relative x_1) is given by:

i) If $h > 1$,

$$Y_1 = A_{02} x_1 + \dots + A_{0h} x_1^{h-1} + Z_1 x_1^{h-1}$$

$$\overline{Z}_{j-1} = \sum_i A_{ji} z_j^i + Z_{j+1} z_j^{h_j} , \quad 1 \leqslant j \leqslant r.$$

ii) If $h = 1$,

$$\overline{Z}_{j-1} = \sum_i A_{ji} z_j^i + Z_{j+1} z_j^{h_j} , \quad 1 \leqslant j \leqslant r.$$

2.4.9.3.- Let

$$\overline{Z}_{j-1} = \sum_i A_{ji} z_j^i + Z_{j+1} z_j^{h_j} , \quad 0 \leqslant j \leqslant r,$$

be a Hamburger-Noether expansion for $\square$ in some of the basis of its maximal ideal. If $n_j = \underline{v}(z_j)$, $0 \leqslant j \leqslant r$, $(x_1 = z_0)$, then

i) The different multiplicities of rings which occur in the desingularization sequence for $\square$ (see 1.5.10.) are the integers n_j, $0 \leqslant j \leqslant r$.

ii) Each n_j is repeated exactly h_j times, $0 \leqslant j \leqslant r$, $(h_0 = h,\ h_r = \infty)$.

iii) The integers h_j and n_j depend only on $\square$.

2.4.9.4.- A Hamburger-Noether expansion determines uniquely a sequence of formal quadratic transformations of N-spaces which desingularize the embedded curve C defined by the choice of the basis $\{x_i\}_{1 \leqslant i \leqslant N}$. Moreover, the sequence of infinitely near points of the origin of C is composed by h points of multiplicity n, h_1 points of multiplicity n_1, etc...

2.4.9.5.- Two curves $\square$ and $\square^*$ are equiresoluble if and only if the integers h_j and n_j of two respective Hamburger-Noether expansions (D) and (D^*) agree.

2.4.9.6.- Any set of type (D) which verifies the requirement on A_{j1} stated in the remark 2.4.5. are a Hamburger-Noether expansi of some curve in some basis of its maximal ideal.

CHAPTER III

CHARACTERISTIC EXPONENTS OF PLANE ALGEBROID CURVES

This chapter is devoted to the study of equisingularity of irreducible algebroid plane curves over an algebraically closed field.

Here the main idea is the construction of a complete system of invariants for the equiresolution. These invariants will be called characteristic exponents. In the characteristic zero case they are computed in the usual way by means of Puiseux expansions. In positive characteristic and in successive sections, we shall compute them by using Hamburger-Noether expansions and Newton polygons, proving, when $(n,p)=1$, they agree with the classical characteristic exponents.

Since the equisingularity will be considered in this chapter, we begin it by giving in the first section, a short account of Zariski's theory of equisingularity for plane curves and its meaning in the case of germs of complex analytic curves.

Throughout all this chapter, the word "curve" will stand for plane curve.

1. REPORT ON EQUISINGULARITY THEORY.

In this section we give, without proofs, the basic facts and definitions in the equisingularity theory initially developed by Zariski [28], and we consider, in particular, the case of complex analytic plane curves.

We begin by considering embedded plane algebroid curves C over an algebraically closed field k. We assume C to be, in general, reducible and denote by $(C) = \{\Gamma_1, \ldots, \Gamma_m\}$ the set of irreducible components of C and by $\Delta(C) = \{t_1, \ldots, t_s\}$ the set of distinct tangents of C. The strict (resp. total) quadratic transform $T_{st}(C) = \bigcup_{j=1}^{s} C'_j$ (resp. $T_{tot}(C) = \bigcup_{j=1}^{s} C^*_j$), introduced in chapter I, 1.5.14. and 1.5.18., has s connected components C'_j (resp. C^*_j).

Definition 3.1.1.- Let C, D two embedded plane curves. A bijection $\pi : (C) \longrightarrow (D)$ is called a pairing between C and D. We shall say that a pairing π is tangentially stable provided that two irreducible components Γ_i and Γ_j of C are tangent iff so are $\pi(\Gamma_i)$ and $\pi(\Gamma_j)$.

Remarks 3.1.2.- 1) Let $\pi : (C) \longrightarrow (D)$ be a tangentially stable pairing. If $\Delta(D)$ is suitably ordered then π induces pairings

$$\left.\begin{array}{l} \pi'_j : (C'_j) \longrightarrow (D'_j) \\ \pi^*_j : (C^*_j) \longrightarrow (D^*_j) \end{array}\right. , \quad 1 \leq j \leq s ,$$

which are not tangentially stable in general.

2) If C and D are irreducible, the trivial pairing $(C) \longrightarrow (D)$ is tangentially stable.

Definition 3.1.3.- A pairing π $(C) \longrightarrow (D)$ is said to be an (a)-equivalence (of singularities) iff either C and D are regular, or π verifies the following conditions:

i) π is tangentially stable.

ii) $e(\Gamma_i) = e(\pi(\Gamma_i))$ for all $\Gamma_i \in (C)$. ($e(-)$ stands for multiplicity as in chapter I.)

iii) The pairings $\pi'_j : (C'_j) \longrightarrow (D'_j)$ induced by π are (a)-equivalences for all j.

The coherence of the inductive method used in the above definition is guaranteed by the fact that by a finite number of successive quadratic transformations we obtain a strict transform of C which has regular curves as connected components, (see 1.5.18).

Remarks 3.1.4.- 1)Two curves are said to be (a)-equisingular when they correspond by an (a)-equivalence.

2) If strict and total transforms are used simultaneously then (a)-equivalence may be characterized without reference to the multiplicity of the irreducible components. Then one may prove that the definition of (a)-equivalence is equivalent to the following one:

We say that a pairing $\pi : (C) \longrightarrow (D)$ is a (b)-equivalence iff either C and D have an ordinary double point or π verifies the following conditions:

i) π is tangentially stable.

ii) The induced pairings $\pi'_j : (C'_j) \longrightarrow (D'_j)$ and $\pi^*_j : (C^*_j) \longrightarrow (D^*_j)$ are (b)-equivalences for all j.

The coherence of the definition is guaranteed, according to 1.5.18., by the fact that by a finite number of quadratic transformations the connected components of the successive total transform of C have only ordinary double points.

3) A third definition, equivalent to the two precedent, which does not make reference to a concrete pairing may be given by means of the formal equivalence. Two curves C and D are said to be formally equivalent if and only if one of the following conditions holds:

a) C and D are regular.

b) C and D have simultaneously an ordinary double point.

c) There exists a tangentially stable pairing $\pi :(C) \longrightarrow (D)$ and an ordering on $\Delta(D)$ such that $\forall\ t_j \in \Delta(C)$, C'_j and D'_j, and C^*_j and D^*_j, are formally equivalent.

4) If C and D are irreducible, clearly they are (a)-equisingular iff they have the same sequence of multiplicities in the

respective desingularization sequences 1.5.10. Therefore, <u>(a)-equisingularity agrees with equiresolution</u>.

In the general case, a pairing $\pi:(C) \longrightarrow (D)$ is an (a)-equivalence if and only if

a) $\forall\, \Gamma_i \in (C)$, Γ_i and $\pi(\Gamma_i)$ are equiresoluble.

b) $\forall\, \Gamma_i, \Gamma_j \in (C)$, $(\Gamma_i, \Gamma_j) = (\pi(\Gamma_i), \pi(\Gamma_j))$, where $(-,-)$ stands for intersection multiplicity.

<u>Thus, the study of (a)-equisingularity for plane algebroid curves is reduced essentially to that of the equiresolution for irreducible algebroid curves.</u>

Henceforth, we shall consider only the case of irreducible curves, and so (a)-equisingularity and equiresolution will be identified. The concept of equiresolution is intrinsic, i.e, it does not depend on embeddings, therefore equiresolution is a weaker relation than formal equivalence for embedded curves. Recall that two embedded curves are said to be formally equivalent over k iff they correspond by a change of variables given by formal series, $x'=f(x,y)$, $y'=g(x,y)$, with Jacobian different from 0, and hence iff its local rings are k-isomorphic. Thus, the multiplicities of the successive quadratic transforms are <u>invariant</u> under formal changes of variables.

In general, a number associated to embedded curves is said to be an <u>invariant</u> when for each curve it depends only on it class modulo formal equivalence, i.e., when it can be constructed from the local ring.

By a <u>complete system of invariants for the equiresolution</u> we mean a family of invariants which determines and is determined by the sequence of multiplicities of the desingularization chain.

<u>Remark 3.1.5.- (Characteristic exponents)</u>. In this remark we consider only irreducible algebroid curves over an algebraically closed field of characteristic zero.

For such a curve $\square$, if $\{x,y\}$ is a basis of its maximal ideal such that x is transversal, we have a primitive Puiseux parametric representation

$$(1)\qquad x = t^n \qquad y = \sum_{i=n}^{\infty} a_i t^i \ .$$

Write $\beta_0 = n$. If $\beta_0 > 1$, we define

$$\beta_1 = \min \{ i \mid a_i \neq 0 \text{ and } (i, \beta_0) < \beta_0 \}.$$

We then proceed inductively. If β_ν is defined and $(\beta_0,\dots,\beta_\nu) > 1$ we set

$$\beta_{\nu+1} = \min\{ i \mid a_i \neq 0 \text{ and } (\beta_0,\dots,\beta_\nu, i) < (\beta_0,\dots,\beta_\nu)\} .$$

Since the greatest commun divisor of the set $\{i \mid a_i \neq 0\}$ is 1 (see 2.1.13), then there exists an integer $g > 0$ such that $(\beta_0,\dots,\beta_g) = 1$. In this situation, the integers $\beta_0,\dots,\beta_g$ are called the <u>characteristic exponents</u> of the parametrization (1). It can be shown that they do not depend on the choice of the basis $\{x,y\}$ nor on the parametrization (Zariski, [28], III), and hence they may be called the <u>characteristic exponents of</u> $\square$.

Moreover, characteristic exponents are a complete system of invariants for equiresolution. This is a consequence of the <u>inversion formula</u> of Abhyankar, [1]:

If in (1) m is the order of y in t, we have another parametric representation

$$(2)\qquad x = \sum_{i=n}^{\infty} b_i t'^i \qquad y = t'^m$$

for a suitable t'. Characteristic exponents $\beta'_0 = n, \beta'_1, \ldots, \beta'_{g'}$ can be defined for (2) in the same way. Then:

a) If $n < m < \beta_1$, then $g' = g+1$, $\beta'_1 = n$ and $\beta'_{\nu+1} = \beta_\nu + n - m$, $1 \leq \nu \leq g$.

b) If $n = m$ or $m = \beta_1$, then $g' = g$ and $\beta'_\nu = \beta_\nu + n - m$, $1 \leq \nu \leq g$.

Remark 3.1.6.-(Characteristic pairs). In the case $k = \mathbb{C}$ a system of pairs of integers, called characteristic pairs, play an important role.

They are constructed by recurrence from the characteristic exponents in the following way: The first pair (m_1, n_1) is given by the conditions $\beta_1/n = m_1/n_1$ and $g.c.d.(m_1, n_1) = 1$. If $i-1$ pairs have been defined then (m_i, n_i) is given by the conditions

$$\frac{\beta_i}{n} = \frac{m_i}{n_1 \cdots n_{i-1} n_i} \quad \text{and} \quad g.c.d.(m_i, n_i) = 1.$$

The set $((m_i, n_i))_{1 \leq i \leq g}$ determines completely the characteristic exponents and verifies $g.c.d.(m_i, n_i) = 1$ and $m_{i-1} n_i < m_i$, $2 \leq i \leq g$. Conversely, any set of pairs with these conditions is the set of characteristic pairs for a curve.

Remark 3.1.7.-(Toroidal knots). Let $F = ((m_i, n_i))_{1 \leq i \leq g}$ be a set of pairs of positive integers with $g.c.d.(m_i, n_i) = 1$, and $(p_i)_{1 \leq i \leq g}$ the set of integers constructed inductively by

$$p_1 = m_1$$
$$p_i = m_i - m_{i-1} n_i + p_{i-1} n_{i-1} n_i, \quad i \geq 2.$$

The image $T^{m_1}_{n_1}$ of the map

$$\begin{aligned} S^1 &\longrightarrow S^1 \times S^1 \qquad (S^1 = \text{unit circle}) \\ z &\longmapsto (z^{m_1}, z^{n_1}) \end{aligned}$$

is called toroidal knot of type (m_1, n_1).

If the toroidal knot of type $(m_1,n_1),\dots,(m_{i-1},n_{i-1})$ $T^{m_1\dots m_{i-1}}_{n_1\dots n_{i-1}}$ is defined, we define $T^{m_1\dots m_i}_{n_1\dots n_i}$ as follows: Take a tubular neighbourhood K_{i-1} of $T^{m_1\dots m_{i-1}}_{n_1\dots n_{i-1}}$ whose bord is homeomorphic to $S^1\times S^1$. Then $T^{m_1\dots m_i}_{n_1\dots n_i}$ is the image of $T^{p_i}_{n_i}$ by that homeomorphism.

A toroidal knot $T^{m_1\dots m_g}_{n_1\dots n_g}$ is called algebraic if for all i, $2\leqslant i\leqslant g$, $p_i > p_{i-1}n_{i-1}n_i$, or equivalently if $m_{i-1}n_i < m_i$, i.e., if F is the set of characteristic pairs for a curve.

<u>Remark 3.1.7.-(Toroidal knot of a complex analytic plane curve singularity)</u>. Let $f(x,y)$ a convergent irreducible power series at the origin $O=(0,0)$ of $\mathbb{C}^2$, and let C the irreducible algebroid curve over $\mathbb{C}$ of equation $f(x,y)=0$.

Taking $\mathbb{C}^2\cong\mathbb{R}^4$, the variety of $\mathbb{C}^2$ of equation $f(x,y)=0$ is identified locally in a neighbourhood of the origin to a surface C_o of $\mathbb{R}^4$. The point O is an isolated singularity of C_o (Milnor, [17]) and there exist $\varepsilon_o>0$ such that $\forall\,\varepsilon$, $0<\varepsilon\leqslant\varepsilon_o$ the sphere S_ε with centre O and radius ε, meet C_o transversaly and the intersection $C_o\cap S_\varepsilon$ is a nonsingular real curve, which by a theorem by Brauner [7] is actually the toroidal knot defined by the characteristic pairs of C.

<u>Remark 3.1.8.-(Topological type of a singularity).</u> The fact, explained in the above remark, that the characteristic pairs of a curve determine the toroidal knot associated with it , is complemented with the following result by Milnor [17]:

<u>Theorem:</u> There exists $\varepsilon_o>0$ such that $\forall\,\varepsilon$, $0<\varepsilon\leq\varepsilon_o$ the pairs of differentiable manifolds $(S_\varepsilon, S_\varepsilon\cap C_o)$ are diffeomorphic. Moreover, if $C(S_\varepsilon\cap C_o)$ is the real projecting cone of $S_\varepsilon\cap C_o$ from O and if B_ε is the closed ball with centre O and radius ε, the pairs of topological spaces $(B_\varepsilon, B_\varepsilon\cap C_o)$ and $(B_\varepsilon, C(S_\varepsilon\cap C_o))$ are homeomorphic.

This theorem establishes therefore that the topological type of $(B_\varepsilon, B_\varepsilon \cap C_o)$ and the isotopy type of $(S_\varepsilon, S_\varepsilon \cap C_o)$ are uniquely determined by the characteristic pairs, and hence by the characteristic exponents of C, thus justifying the adjetive characteristic.

Conversely, results by Brauner, [7], Burau, [8] and Zariski, [27] show that the characteristic pairs of an analytically irreducible plane curve depend only on the topology of the singularity.

In the case of reducible curves, we have the following result, (Lejeune, [15], and Zariski, [27]):

Theorem: The topological type of a singularity of a complex analytic plane curve is determined by the topological type of each analytically irreducible component and the intersection multiplicity of any pairs of distinct components.

Remark 3.1.9.- (Moduli of local singularities). One may consider the problem of classifying, modulo analytic isomorphism, all complex analytic singularities with a given embedding topological type. This problem, which is analogous to the Riemann moduli problem for global curves, may be studied by means of purely algebraic techniques, (see Zariski (26)). In particular, it is equivalent to the problem of classifying modulo formal equivalence all algebroid plane curves with a given (a)-equisingularity class. The word "invariant" introduced in 3.1.4. may be interpreted easily in terms of this moduli problem.

2. CHARACTERISTIC EXPONENTS.

Consider an irreducible plane algebroid curve $\square$ over an algebraically closed field k.

Definition 3.2.1.- We shall define the genus of a curve $\square$ by recurrence on the number of quadratic transformations needed to desingularize $\square$, in the following way:

Let $\square_1$ be the quadratic transformed of $\square$,

(i) If $\square$ is regular: $g(\square) = 0$.

(ii) If $e(\square) > e(\square_1)$,and $e(\square_1)$ does not divide $e(\square)$: $g(\square) = g(\square_1)$.

(iii) If $e(\square) > e(\square_1)$, and $e(\square_1)$ divides $e(\square)$: $g(\square) = g(\square_1) + 1$.

(iv) If $e(\square) = e(\square_1)$: $g(\square) = g(\square_1)$.

<u>Remark 3.2.2.</u>- From the above definition $g(\square) = 0$ iff $\square$ is regular, and using the Hamburger-Noether expansion, the genus of the curve can be trivially computed by the formula:

$$g(\square) = \# \left(\left\{ \frac{n}{n_1}, \frac{n_1}{n_2}, \dots, \frac{n_{r-1}}{n_r} \right\} \cap Z \right) .$$

<u>Definition 3.2.3.</u>- Consider a plane algebroid curve $\square$. We define a <u>complex model for $\square$</u> to be a plane curve $\square_{\mathbb{C}}$ over the complex field $\mathbb{C}$ having the same singularity reduction process (1.5.10) as $\square$.

More precisely, the condition requires that if

$$\square \subset \square_1 \subset \dots \subset \square_M = \overline{\square}$$
$$\square_{\mathbb{C}} \subset (\square_{\mathbb{C}})_1 \subset \dots \subset (\square_{\mathbb{C}})_{M'} = \overline{\square}_{\mathbb{C}}$$

are the respective desingularization sequences for $\square$ and $\square_{\mathbb{C}}$, then:

(i) $M = M'$.

(ii) $e(\square_i) = e((\square_{\mathbb{C}})_i)$, $0 \leqslant i \leqslant M$.

<u>Proposition 3.2.4.</u>- Let $\square$ be a curve with a Hamburger-Noether expansion given by:

$$(\underline{D}) \qquad z_{j-1} = \sum_i a_{ji} z_j^i + z_j^{h_j} z_{j+1} \; , \quad 0 \leq j \leq r,$$

and denote by $F : k \longrightarrow \mathbb{C}$ any map verifying $F(x) \neq 0 \iff x \neq 0$. Then the complex curve $\square_{\mathbb{C}}$ which has

$$(\underline{D}_{\mathbb{C}}) \qquad z'_{j-1} = \sum_i F(a_{ji}) z_j'^{\,i} + z_j'^{\,h_j} z'_{j+1} \; , \quad 0 \leq j \leq r,$$

as Hamburger-Noether expansion is a complex model for $\square$.

Proof: First, recall that such a curve with expansion $(\underline{D}_{\mathbb{C}})$ exists, (see 2.2.13.).

In order to prove the proposition notice that it suffices to check that the values h_j and n_j in both expansions $(\underline{D})$ and $(\underline{D}_{\mathbb{C}})$ agree (2.2.10.).

By construction r and the h_j coincide. If $n'_j = \underline{v}(z'_j)$, we have $n_r = n'_r = 1$, and by looking at the requirement on F, it follows trivially that $n_j = n'_j$, $0 \leq j \leq r$.

Proposition 3.2.5.- Let $\square$ (resp. $\square^*$) a curve over k, and assume that $\square_{\mathbb{C}}$ (resp. $\square^*_{\mathbb{C}}$) is a complex model for $\square$ (resp. $\square^*$). The curves $\square$ and $\square^*$ are (a)-equisingular if and only if the complex curves $\square_{\mathbb{C}}$ and $\square^*_{\mathbb{C}}$ are (a)-equisingular.

Proof: Since the desingularization sequences for $\square$ and $\square_{\mathbb{C}}$ (resp. $\square^*$ and $\square^*_{\mathbb{C}}$) agree the proof is evident.

Corollary 3.2.6.- All the complex models for a curve $\square$ are (a)-equisingular.

Proposition 3.2.7.- If $\square_{\mathbb{C}}$ is a complex model for $\square$, then $g(\square) = g(\square_{\mathbb{C}})$.

Proof: Let n_j, $0 \leq j \leq r$, be the different multiplicities which occur in the desingularization sequence for $\square$ (actually for $\square_{\mathbb{C}}$). Then we have:

$$g(\square) = \#\left(\left\{\frac{n_o}{n_1}, \ldots, \frac{n_{r-1}}{n_r}\right\} \cap \mathbb{Z}\right) = g(\square_{\mathbb{C}}).$$

Definition 3.2.8.- Consider a curve □ over the algebraically closed field k, and assume that $\square_{\mathbb{C}}$ is a complex model for □. We define the characteristic exponents of □ to be the characteristic exponents of the complex curve $\square_{\mathbb{C}}$.

Remark 3.2.9.- By 3.2.4. there exist complex models. By 3.2.5. all of them are (a)-equisingular. Then the characteristic exponents do not depends on the complex model. Furthermore, by 3.2.7. the number of characteristic exponents is $g(\square)+1$.

The above definition is therefore coherent, and the characteristic exponents depend only on the ring □.

We remark also that the characteristic exponents are non-negative integers $(\beta_\nu)_{0 \leq \nu \leq g}$ such that:

(i) $e(\square) = \beta_0 < \beta_1 < \ldots < \beta_g$.

(ii) $\beta_0 > (\beta_0, \beta_1) > \ldots > (\beta_0, \ldots, \beta_g) = 1$.

Proposition 3.2.10.- Two curves are (a)- equisingular if and only if they have the same genus and the same set of characteristic exponents. In other words, we may state the following: "The set of characteristic exponents is a complete system of invariants for the (a)-equisingularity".

Lemma 3.2.11.- Let □ be a curve and denote by $\square_1$ its quadratic transform. Let $(\beta_\nu)_{0 \leq \nu \leq g}$ (resp. $(\beta'_\nu)_{0 \leq \nu \leq g'}$) be the characteristic exponents of □ (resp. $\square_1$). If $\beta_1 = hn + n_1$, where $0 < n_1 < n$, the following statements hold:

(a) If $h > 1$,

$g = g'$, $\beta_0 = \beta'_0$, $\beta'_\nu = \beta_\nu - n$, $1 \leq \nu \leq g$.

(b) If $h = 1$, and n_1 does not divide n,

$g = g'$, $\beta'_0 = n_1$, $\beta'_\nu = \beta_\nu - n_1$, $1 \leqslant \nu \leqslant g$.

(c) If $h = 1$, and n_1 divides n,

$g = g'+1$, $\beta_0 = n_1$, $\beta'_\nu = \beta_{\nu+1} - n_1$, $1 \leqslant \nu \leqslant g-1$.

Proof: If $\square_{\mathbb{C}}$ is a complex model for $\square$, then $(\square_{\mathbb{C}})_1$ is a complex model for $\square_1$. Consequently, it is sufficient to prove the result only in the case in which k is the complex field.

But, in this case, any curve may be taken as a complex model for itself, and so the proposition derives from the "inversion formula" (see 3.1.5.).

Remark 3.2.12.- This inversion formula, may be used too, when k has characteristic zero, to find relations, as in the above lemma, among the characteristic exponents given by Puiseux series for any curve and its quadratic transform. This leads, in particular, to the following result:

Proposition 3.2.13.- Let k be a field of characteristic 0. The characteristic exponents of any curve over k agree with the characteristic exponents of any Puiseux series representing the curve.

Proof: By induction on $M = M(\square)$ (the minimum number of quadratic transformations needed to desingularize $\square$).

For $M = 0$, the proof is evident. Let $\square$ be a curve over k, for which $M = M(\square) > 0$, and assume that the proposition is proved for each $\square'$ with $M(\square') < M$. Denote by $(\beta_\nu)_{0 \leqslant \nu \leqslant g}$ (resp. by $(\beta'_\nu)_{0 \leqslant \nu \leqslant g'}$) the characteristic exponents for $\square$ (resp. the quadratic transform $\square_1$). The same denote by $(\beta^*_\nu)_{0 \leqslant \nu \leqslant g^*}$ (resp. $(\beta'^*_\nu)_{0 \leqslant \nu \leqslant g'^*}$) the characteristic exponents of the Puiseux series which represent $\square$ (resp. $\square_1$).

By the induction hypothesis $g' = g'^{*}$ and $\beta'_{\nu} = \beta'^{*}_{\nu}$, $0 \leqslant \nu \leqslant g'$. On the other hand, it is evident that $\beta_0 = \beta_0^{*}$ and $\beta_1 = \beta_1^{*}$, hence, using the above lemma and the remark 3.2.12., we deduce $g = g^{*}$ and $\beta_{\nu} = \beta_{\nu}^{*}$ for each ν, $0 \leqslant \nu \leqslant g$.

3. CHARACTERISTIC EXPONENTS AND HAMBURGER-NOETHER EXPANSIONS.

Notations 3.3.1.- Consider the irreducible plane algebroid curve $\square$ over an algebraically closed field k. Suppose that a basis of the maximal ideal $\underline{m}$ is chosen and denote by

$$(\underline{D}) \qquad z_{j-1} = \sum_i a_{ji} z_j^{\,i} + z_j^{h_j} z_{j+1}, \quad 0 \leqslant j \leqslant r,$$

the Hamburger-Noether expansion for $\square$ in that basis $(x = z_o, \ y = z_{-1})$.

If $\underline{v}$ is the associated valuation to $\square$, and we set $n_j = \underline{v}(z_j)$, $0 \leqslant j \leqslant r$; the integers h_j, n_j, $0 \leqslant j \leqslant r$, $(h_o = h,\ h_r = \infty)$, are invariants of the curve (see proposition 2.2.10.).

Let $s_1 < s_2 < \dots < s_g$ be the ordered set of the indices j, $0 \leqslant j \leqslant r$, for which $n_j \nmid n_{j-1}$. For convenience of notations we shall also put $s_o = 0$. Notice that g is exactly the genus of $\square$.

Proposition 3.3.2.- Let $\square$ be an algebroid curve, whose characteristic exponents are $(\beta_{\nu})_{0 \leqslant \nu \leqslant g}$. Then:

$$\beta_0 = n_0 = n,$$

$$\beta_{\nu+1} = \sum_{j=0}^{s_{\nu}} h_j n_j + n_{s_{\nu}} + n_{s_{\nu+1}} - n.$$

Proof: We use again induction on $M(\square) = M$. For $M = 0$ the result is evident.

Let $\square$ be a curve, with characteristic exponents $(\beta_\nu)_{0\leqslant\nu\leqslant g}$, and let $(\underline{D})$ be a Hamburger-Noether expansion in a certain basis. Let $\square_1$ be the quadratic transform of $\square$, and $(\beta'_\nu)_{0\leqslant\nu\leqslant g'}$ its characteristic exponents.

We shall consider three cases as in lemma 3.2.11.:

(a) $h>1$. Actually $g'=g$, $\beta'_0=\beta_0$, $\beta'_\nu=\beta_\nu-n$, $1\leqslant\nu\leqslant g$. By the induction hypothesis, applied to $\square_1$, and the above relations we obtain the expresions for β_ν as desired.

(b) $h=1$ and $n_1 \nmid n$. We have $g'=g$, $\beta'_0=n_1$, $\beta'_\nu=\beta_\nu-n_1$, $1\leqslant\nu\leqslant g$. The proof follows as in (a).

(c) $h=1$ and $n_1 \mid n$. We have $g'=g-1$, $\beta'_0=n_1$, $\beta'_\nu=\beta_{\nu+1}-n_1$, $1\leqslant\nu\leqslant g-1$. But in this case $s_1=1$.

It is trivial that $\beta_0=n$ and $\beta_1=n+n_1$.

For $\nu>1$, applying the induction hypothesis to $\square_1$, we get:

$$\beta_{\nu+1}=n_1+\beta'_\nu=n_1+\sum_{j=1}^{s_\nu} h_j n_j + n_{s_\nu} + n_{s_{\nu+1}} - n_1 =$$

$$=\sum_{j=0}^{s_\nu} h_j n_j + n_{s_\nu} + n_{s_{\nu+1}} - n.$$

<u>Proposition 3.3.3.-</u> Let $\square$ be a curve with Hamburger-Noether expansion given by

$$z_{j-1}=\sum_i a_{ji} z_j^i + z_j^{h_j} z_{j+1}, \quad 0\leqslant j\leqslant r.$$

Keeping the notations as in the above proposition, the following statements hold:

(a) If $j\neq s_\nu$, $0\leqslant\nu\leqslant g$, then $a_{ji}=0$, $1\leqslant i\leqslant h_j$.

Furthermore, in this case $n_{j-1}=h_j n_j + n_{j+1}$.

(b) If $j = s_\nu$, $1 \leq \nu \leq g$, there exists an integer $k_\nu \leq h_{s_\nu}$ such that $a_{ji} = 0$, $1 \leq i < k_\nu$, and $a_{jk_\nu} \neq 0$. In this case $n_{j-1} = k_\nu n_j$.

(c) $(\beta_0, \beta_1, \ldots, \beta_\nu) = (n_{s_{\nu-1}}, n_{s_{\nu-1}+1}) = n_{s_\nu}$, $1 \leq \nu \leq g$.

Proof: (a) and (b) proceed directly from the properties of the indices s_ν.

We shall prove (c) by using induction on ν. For $\nu = 1$, by (a) and (b) n_{s_1} is the last remainder in the Euclid's algorithm for n and n_1, so it is its greater common divisor. Thus,

$$(\beta_0, \beta_1) = (n, n_1) = n_{s_1} .$$

Now, if (c) holds for ν, then $(\beta_0, \ldots, \beta_\nu) = (n_{s_\nu}, \beta_{\nu+1})$. By using a similar argument as in the case $\nu = 1$ and according to 3.3.2. we obtain

$$(n_{s_\nu}, \beta_{\nu+1}) = (n_{s_\nu}, n_{s_\nu+1}) = n_{s_{\nu+1}} .$$

Remark 3.3.4.- Keeping the notations as in the preceding proposition, the Hamburger-Noether expansion of a curve □ in a basis $\{x, y\}$ may be written in the following way:

$$\text{(D')} \quad \begin{aligned} y &= a_{01} x + \ldots + a_{0h} x^h + x^h z_1 \\ x &= z_1^{h_1} z_2 \\ &\ldots\ldots\ldots\ldots\ldots\ldots \\ z_{s_1-1} &= a_{s_1 k_1} z_{s_1}^{k_1} + \ldots + a_{s_1, h_{s_1}} z_{s_1}^{h_{s_1}} + z_{s_1}^{h_{s_1}} z_{s_1+1} \\ z_{s_1} &= z_{s_1+1}^{h_{s_1+1}} z_{s_1+2} \\ &\ldots\ldots\ldots\ldots\ldots\ldots\ldots\ldots \\ z_{s_g-1} &= a_{s_g k_g} z_{s_g}^{k_g} + \ldots . \end{aligned}$$

with $a_{s_\nu k_\nu} \neq 0$, $1 \leq \nu \leq g$.

Notice that the integers h_j , $0 \leq j \leq r = s_g$, and k_ν , $1 \leq \nu \leq g$, determine trivially the (a)-equisingularity class of □.

Thereafter, unless otherwise specified, we shall keep the notations which are been used here. When a Hamburger-Noether expansion (D) should be considered, we shall assume that this expansion is actually written in the reduced form (D').

Remark 3.3.5.- The reduced form (D') of the Hamburger-Noether expansion provides complete information about the infinitely near points of the origin O of the curve (see 1.5.15.-1.5.20.):

(A) Free points.

1. The h points of multiplicity n and the first one of multiplicity n_1.

2. The $h_{s_\nu} - k_\nu$ last points of multiplicity n_{s_ν} , and the first one of multiplicity $n_{s_\nu + 1}$, $(1 \leq \nu \leq g)$. Here we may suppose that $h_{s_g} = \infty$.

Note that there exists a one-one correspondence from the free infinitely near points onto the coefficients in (D') which are not necessarily zero.

(B) Satellite points.

All the points which are not included in the above list of free points.

(C) Leading free points.

They are the following g points:

$$O'_\nu = O_{h+h_1+\ldots+h_{s_{\nu-1}}+k_\nu} , \quad 1 \leq \nu \leq g.$$

(D) Terminal satellite points.

They are the following g points:

$$O^*_\nu = O_{h+h_1+\dots+h_{s_\nu -1}+k_\nu -1}, \qquad 1 \leq \nu \leq g.$$

(E) Proximate points of O_i.

1. If $e(O_i) = e(O_{i+1})$, O_{i+1} is the only proximate point of O_i.

2. If $n_{j-1} = e(O_i) > e(O_{i+1}) = n_j$, then:

(i) If $j \neq s_\nu$, $1 \leq \nu \leq g$, the proximate points of O_i are the h_j points of multiplicity n_j and the first one of multiplicity n_{j+1}, (here $h_{s_g} = \infty$).

(ii) If $j = s_\nu$ for some $\nu, 1 \leq \nu \leq g$, the proximate points O_i are the k_ν first points of multiplicity $n_j = n_{s_\nu}$.

We remark that the classical results on satellitisme and proximity (which may be found for instance in Zariski, (27), GTS, II, 5), for infinitely near points of the origin of an irreducible curve can be obtained trivially from this analysis.

Lemma 3.3.6.- Consider a rational number m/n, $m > n > 0$. Assume that

$$\begin{aligned} m &= c_0 n + r_1 \\ n &= c_1 r_1 + r_2 \\ &\dots\dots\dots\dots \\ r_{s-1} &= c_s r_s \end{aligned}$$

denotes the Euclid's algorithm for the integers m and n. Then

$$\sum_{j=0}^{s} c_j r_j + r_s - n = m, \qquad (r_0 = n).$$

Proof: We shall use induction on s.

For $s=1$, it is evident. Suppose that for $s-1$ the formula holds. Applying the induction hypothesis to the rational number n/r_1 we obtain

$$\sum_{j=0}^{s} c_j r_j + r_s - n = c_0 n + r_1 + \sum_{j=1}^{s} c_j r_j + r_s - r_1 - n = m .$$

Proposition 3.3.7.- For a curve, with characteristic exponents $(\beta_\nu)_{0\leqslant \nu\leqslant g}$, the following equalities hold:

$$\beta_0 = n \ ; \qquad \beta_1 = hn + n_1 \ ;$$

$$\beta_{\nu+1} = \beta_\nu + (h_{s_\nu} - k_\nu)\, n_{s_\nu} + n_{s_\nu+1} \ , \quad 1\leqslant \nu \leqslant g-1 .$$

Proof: For β_0, β_1 the corresponding equalities are evident. Now, suppose $\nu > 1$. Using 3.3.2. we have

$$\beta_{\nu+1} - \beta_\nu = -n_{s_\nu} + \Big[(h_{s_{\nu-1}+1} - 1)n_{s_{\nu-1}+1} + \ldots + (k_\nu - 1)n_{s_\nu}\Big] +$$

$$+ (h_{s_\nu} - k_\nu)\, n_{s_\nu} + n_{s_\nu+1} .$$

By the preceding lemma the bracket has the value n_{s_ν} and this completes the proof.

Remark 3.3.8.- The characteristic exponents can be expressed also using infinitely near points as follows:

If $(O^*_\nu)_{0\leqslant\nu\leqslant g}$ denotes the set of terminal satellite points in the sequence of infinitely near points of the origin of the curve, then g is exactly the genus and we have

$$\beta_\nu = \sum_{i\in I_\nu} e(O_i) \quad , \quad 1\leqslant \nu \leqslant g,$$

where I_ν denotes the set formed by the indices i for which O_i is a free point anterior to O^*_ν.

Remark 3.3.9.- Proposition 3.3.2. and 3.3.7. give us directly expresions of the characteristic exponents in function of the Hamburger-Noether expansion.

Conversely, if we know the characteristic exponents of a curve $(\beta_\nu)_{0\leq\nu\leq g}$, we may calculate the integers h_j, n_j and k_ν of the Hamburger-Noether expansion in the following way:

First,

$$n_{s_\nu} = (\beta_0, \ldots, \beta_\nu) \quad , \quad 0 \leq \nu \leq g.$$

(1)

$$h_{s_\nu} - k_\nu = \left[\frac{\beta_{\nu+1} - \beta_\nu}{n_{s_\nu}} \right] \quad , \quad 1 \leq \nu \leq g.$$

Now, we may get $n_{s_{\nu+1}}$ from the equality

$$n_{s_{\nu+1}} = \beta_{\nu+1} - \beta_\nu - (h_{s_\nu} - k_\nu) n_{s_\nu} .$$

Finally, by expressing $n_{s_\nu}/n_{s_{\nu+1}}$ as a continued fraction, one finds the rest of values unless the h_{s_ν} .

But actually the numbers k_ν are already known, then the h_{s_ν} may be computed from the second formula in (1).

Notice that the values h_j, n_j, and k_ν are not of course independent. The characteristic exponents (hence the (a)-equisingularity class) are determined by the number r, by the integers h_j , $0\leq j \leq r$, by the subset of indices $s_1, \ldots, s_g$, and by the integers k_ν , $1 \leq \nu \leq g$, satisfying the condition $k_\nu \leq h_{s_\nu}$. These parameters are certainly independent.

4. CHARACTERISTIC EXPONENTS AND THE NEWTON POLYGON.

We give in the present section an algorithm which enables us to compute the characteristic exponents of a plane curve directly from any of its equations.

Let $\square$ be an irreducible curve and $\{x,y\}$ a basis of $\underline{m}$ with x transversal. Take an equation $f(X,Y)=0$ of the curve in that basis and denote by $O^*_1,\ldots,O^*_g$ the terminal satellite infinitely near points of the origin of the embedded curve. By 1.5.14., 2.2.8 and 2.2.12., f and the transversal parameter x determine uniquely series $f^*_1,\ldots,f^*_g$ defining the successive quadratic transforms $\square^*_1,\ldots,\square^*_g$ of $\square$ at these points.

Using 3.3.7., the first characteristic exponent of $\square^*_\nu$ is given, as function of the characteristic exponents $(\beta_\nu)_{0\leqslant\nu\leqslant g}$ of $\square$, by

$$\beta_1(\square^*_\nu) = (h_{s_\nu} - k_\nu + 1)\, n_{s_\nu} + n_{s_\nu+1} = \beta_{\nu+1} - \beta_\nu + n_{s_\nu} .$$

Thus if $\beta_0,\ldots,\beta_\nu$ and $\beta_1(\square^*_\nu)$ are known, then $\beta_{\nu+1}$ may be computed by the above formula.

We shall get the exponents β_ν successively. For that we must solve two questions: the first one consists in giving a method in order to find the first characteristic exponent of a curve from the Newton polygon of any of its equations, the second one to obtain the Newton diagram of f^*_ν from the Newton diagram of $f^*_{\nu-1}$. Within this section this two questions will be solved.

<u>Definitions and notations</u> 3.4.1.- Let us consider an irreducible plane algebroid curve $\square$, and a basis $\{x,y\}$ of its maximal ideal. By 1.1.3. there is an induced isomorphism

$$k((X,Y))/_{\underline{p}} \cong \square .$$

where the prime ideal $\underline{p}$ is principal. Let $f(X,Y) = \sum_{\alpha,\beta\geq 0} A_{\alpha\beta} X^{\alpha} Y^{\beta}$ be a generator of $\underline{p}$. The series f is determined upon a unit in $k[[X,Y]]$.

If $\mathbf{Z}_+$ (resp. $\mathbf{R}_+$) denotes the set of nonnegative integer (resp. real) numbers, the set

$$D(f) = \{ (\alpha,\beta) \in \mathbf{Z}_+^2 \ / \ A_{\alpha,\beta} \neq 0 \}$$

will be called <u>Newton diagram for f.</u> If the value (or "mass") $A_{\alpha,\beta} \in k$ is attached to $(\alpha,\beta) \in D(f)$, we shall say that it is the <u>weighted Newton diagram of f.</u>

By definition the <u>Newton polygon of f</u> is the set consisting of the straight segments and half lines in the boundary of the convex hull of $D(f) + \mathbf{R}_+^2$ which are not contained in the coordinate axes.

<u>Remark 3.4.2.</u>- The Newton polygon does not depend on the generator of the ideal $\underline{p}$ chosen.

In the particular case $\underline{p} = (Y)$ the Newton polygon is given by $\beta = 1$, $\alpha \geq 0$. The same, if $\underline{p} = (X)$, it is given by $\alpha = 1$, $\beta \geq 0$.

Excluding these trivial cases and since f is irreducible, we may assert that there are points $A = (m,0)$, $B = (0,n)$ in $D(f)$ lying respectively on the coordinate axes. We shall assume that m and n are chosen respectively to be minimal.

Moreover, if $\underline{p} = (Y)$, we shall set $n=1$, $m=\infty$; and if $\underline{p} = (X)$, $n = \infty$, $m = 1$. We shall also assume that $x = X + (f)$ is a transversal parameter for $\square$, (i.e. X does not divide the leading form of f).

<u>Lemma 3.4.3.</u>- If $A = (m,0)$ and $B = (0,n)$ are as above, then $\underline{v}(x) = n$ and $\underline{v}(y) = m$. ($\underline{v}$ being the natural valuation associated to $\square$).

Proof: Let C' be the curve $y = 0$. C' has $x = x$, $y = 0$, as a parametric representation, therefore x is a uniformizing parameter for C'. If C is our curve, we have

$$\underline{v}(y) = (C', C) = (C, C') = \underline{v}_x(f(x,0)) = m .$$

An identical argument holds for $x = 0$.

Lemma 3.4.4.- If $\underline{p} \neq (Y)$, the Newton polygon for f is the straight line segment which joins the points A and B. (Note that $\underline{p} \neq (X)$ since x is transversal).

Proof: We shall prove it by induction on $M(\square)$.

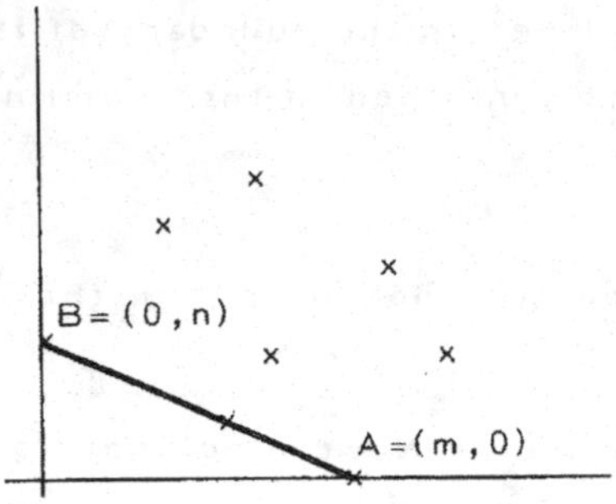

If $M(\square) = 0$, the curve is regular. The result is evident because actually $B = (0,1)$.

Let $\square$ be a curve with $M(\square) > 0$, and assume that for every curve $\square'$ with $M(\square') < M(\square)$ the lemma is true. We consider two cases:

1st case: $m > n$. We refer the quadratic transform $\square_1$ to the basis $\{x, y/x\}$ of its maximal ideal. An equation $f_1 = 0$ of $\square_1$ in this basis is obtained from $f=0$ when the weighted Newton diagram is subject to the plane transfromation

$$T : (\alpha,\beta) \longrightarrow (\alpha+\beta-n, \beta).$$

Moreover the Newton polygons of f and f_1 are homologous by this transformation. Since the polygon of f_1 is a straight line segment, it is the same for that of f. This segment must be necessarily $\overline{AB}$.

2nd case: $m=n$. Since the multiplicity of $\square$ is n, f is a power series of order n. It follows that there are no points (α, β) in the Newton diagram with $\alpha+\beta < n$. Hence the Newton polygon is $\overline{AB}$.

Lemma 3.4.5.- Let f, A, B, m, and n as above. Let P be the Newton polygon of f, and

$$L(X,Y,f) = \sum_{(\alpha,\beta)\in P} A_{\alpha,\beta} X^{\alpha} Y^{\beta} .$$

Then:

(a) If $m=\infty$, $L(X,Y,f) = Y\,U(X,Y)$, where $U(X,Y)$ is a unit in $k[[X,Y]]$.

(b) If $m \neq \infty$ and $m/n = m'/n'$ with $(m',n')=1$, there exist $a, \lambda \in k$, $a \neq 0$, $\lambda \neq 0$, such that

$$L(X,Y,f) = a\,(Y^{n'} - \lambda X^{m'})^{n/n'} .$$

Proof: (a) is evident. We shall prove (b) by using again induction on $M(\square)$. For $M(\square)=0$, the Newton polygon joins the points $A=(m,0)$ and $B=(0,1)$, then

$$L(X,Y,f) = a\,(Y-\lambda X^{m}) , \quad a\neq 0 , \quad \lambda \neq 0.$$

Now, take a curve $\square$ with $M(\square)>0$, and assume that for $M(\square') < M(\square)$ (b) holds. As above we consider two cases:

1st case: $m>n$. The plane transformation T in the preceding lemma maps the Newton polygon of f onto that of f_1.

If $(m,n)=r$, then $(m-n,n)=r$; thus by the induction hypothesis:

$$L(X, \frac{Y}{X}, f_1) = (c' \frac{Y^{n'}}{X^{n'}} - a X^{m'-n'})^r , \quad a \neq 0, \; c' \neq 0.$$

Hence,

$$L(X, Y, f) = X^n L(X, \frac{Y}{X}, f_1) = (c' Y'^n - a X'^m)^r .$$

2nd case: $m = n$. $L(X,Y,f)$ is the leading form of f, then it is of type $a(Y - \lambda X)^n$, $a \neq 0$, since f is irreducible. As $m = n$, we must also have $\lambda \neq 0$.

Proposition 3.4.6.- With notations as in the above lemmas, assume that $\underline{v}(y) = m = nd$, with $d > 0$, $d \in \mathbb{Z}$, and $L(X,Y,f) = a(Y - \lambda X^d)^n$, with $a \neq 0$, $\lambda \neq 0$. Then, if $\bar{y} = y - \lambda x^d$, we have $\underline{v}(\bar{y}) > dn$.

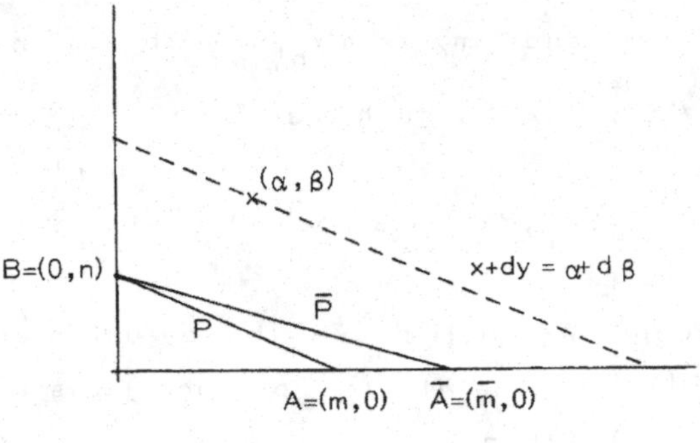

Particulary, by the change $\{x, y\} \longrightarrow \{x, \bar{y}\}$ the Newton polygon is displaced to the right side as in the picture.

Proof: The change $\{x, y\} \longrightarrow \{x, \bar{y}\}$ induces a multivaluated function between the respective Newton diagrams. If $(\alpha, \beta) \in D(f)$,

$$X^\alpha Y^\beta = X^\alpha (\bar{Y} + \lambda X^d)^\beta = X^\alpha \left(\sum_{i=0}^{\beta} \binom{\beta}{i} \lambda^i X^{di} \bar{Y}^{\beta - i} \right),$$

then the point (α, β) is transformed into the points $(\alpha + di, \beta - i)$, $0 \leq i \leq \beta$. All lie on the straight line $x + dy = \alpha + d\beta$, i.e., on the parallel line to P which passes through (α, β).

As a consequence, the new Newton diagram remains in the plane region $x+dy \geq dn$, $x \geq 0$, $y \geq 0$. Thus the polygon $\bar{P}$ joins the points $B = (0,n)$ and $\bar{A} = (\bar{m},0)$, with $\bar{m} \geq dn$. Since $\underline{v}(\bar{y}) = \bar{m}$, to complete the proof it suffices to prove that $\bar{m} > dn$.

But this is evident, because after the change the point $(dn,0)$ does not belong to the diagram.

<u>Corollary 3.4.7.</u>- Let $\square$ be an irreducible plane curve defined by the series $f(X,Y) \in k((X,Y))$. Assume that the Newton polygon of f joins the points $B = (0,n)$ and $A = (m,0)$, with $n \leq m$. Apply successively changes as in 3.4.6. whenever n divides m. Then:

(a) If $\square$ is regular, we get a change of type

$$y^* = y - \sum_{i=1}^{\infty} \lambda_i x^{d_i}, \qquad (d_i < d_{i+1}),$$

for which the curve has $y^* = 0$ as a new equation.

(b) If $\square$ is not regular, after a finite number of steps we get a change of type

$$y^* = y - \sum_{i=1}^{s} \lambda_i x^{d_i}, \qquad (d_i < d_{i+1}),$$

such that if $\underline{v}(y^*) = m^*$ one has $(m^*,n) < n$. Actually m^* is the first characteristic exponent of $\square$.

<u>Proof:</u> (a) It is evident since $n=1$.

(b) Let β_1 be the first characteristic exponent of $\square$. By looking at the Hamburger-Noether expansion of $\square$ in the basis $\{x,y\}$ one sees that $m = \underline{v}(y) \leq \beta_1$. Thus all the possible values of m after changes as in 3.4.6. are bounded by β_1.

It follows that there is a change

$$y^{*} = y - \sum_{i=0}^{s} \lambda_i x^{d_i} \qquad (d_i < d_{i+1}) ,$$

such that n does not divide $m^{*} = \underline{v}(y^{*})$. By looking again at the Hamburger-Noether expansion of □ (now in the basis $\{x, y^{*}\}$) one can see that $m^{*} = \beta_1$.

Remark 3.4.8.- The value $\bar{m}$ (in proposition 3.4.6.) may be computed directly from the Newton diagram without any substitution. Indeed, for every $\gamma > dn$ the points on the straight line $L_\gamma : \alpha + d\beta = \gamma$ are the only points which may be transformed into $(\gamma, 0)$.

The associated mass $\bar{A}_{(\gamma,0)}$ in $(\gamma, 0)$ in the weighted diagram after change is given by

$$\bar{A}_{(\gamma,0)} = \sum_{(\alpha,\beta) \in L_\gamma} A_{(\alpha,\beta)} \lambda^{\beta} .$$

Thus, $\bar{m}$ must be the least integer γ for which $\bar{A}_{(\gamma,0)} \neq 0$.

In the same way, the associated mass $\bar{A}_{(\alpha_1,\beta_1)}$ in any point $(\alpha_1, \beta_1) \in L_\gamma$ is giving by the formula

$$\bar{A}_{(\alpha_1,\beta_1)} = \sum_{\substack{(\alpha,\beta) \in L_\gamma \\ \beta \geq \beta_1}} \lambda^{\beta-\beta_1} \binom{\beta}{\beta_1} A_{(\alpha,\beta)} .$$

We have just solved the first question outlined at the beginning of this section: to find the first characteristic exponent of any curve from the Newton diagram.

Notations 3.4.9.- With the assumptions as in the beginning of the section, we shall study now the relation between f and f_1^{*}. The series $f_1^{*} = f_1^{*}(u,w)$ has order $e_1 = n_{s_1}$ and its leading form is $c(w - a_{s_1k_1}u)^{e_1}$, where $c \neq 0$ and n_{s_1}, $a_{s_1k_1}$ are given by the Hamburger-Noether expansion. We shall use in the sequel the auxiliar series f_1' defined by $f_1'(u,v) = f_1^{*}(u,uv)/u^{e_1}$.

For the sake of simplicity we shall find a relation between

the Newton diagrams of f and f'_1 instead of those of f and f^*_1. Notice that the relation between f'_1 and f^*_1 is trivial.

Assume now that the curve □ is not regular, and that the series f is chosen such that if $x = X + (f)$, $y = Y + (f)$, then $\underline{v}(x) = n$, $\underline{v}(y) = m = \beta_1$ (where β_1 is the first characteristic exponent of □).

The Newton polygon of f is thus a straight line segment joining the points $A = (m,0)$ and $B = (0,n)$.

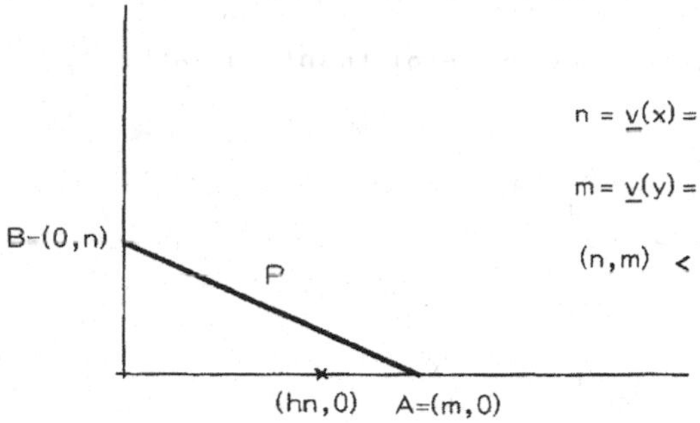

$n = \underline{v}(x) = \beta_0$.

$m = \underline{v}(y) = \beta_1$.

$(n,m) < n$.

The s_1+1 top rows in the Hamburger-Noether expansion of □ in that basis (using the notations as in section 2) are

$$y = x^h z_1$$
$$x = z_1^{h_1} z_2$$
$$\cdots\cdots\cdots$$
$$z_{s_1-2} = z_{s_1-1}^{h_{s_1-1}} z_{s_1}$$
$$z_{s_1-1} = a_{s_1 k_1} z_{s_1}^{k_1} + \ldots, \qquad a_{s_1 k_1} \neq 0.$$

By 3.3.3. the integers $h, h_1, \ldots, h_{s_1-1}, k_1$ are determined by m and n. Namely, the continued fraction which represents the rational fraction m/n has these values as partial quotients:

$$\frac{m}{n} = h + \cfrac{1}{h_1 + \cfrac{1}{h_2 + \cfrac{}{\ddots + \cfrac{1}{k_1}}}} = \frac{[h, h_1, \ldots, k_1]}{[h_1, \ldots, k_1]} .$$

where the brackets [ ] are the well known Euler polynomials, i.e., the polynomials in the partial quotients whose values give respectively the numerator and denominator of the fraction, when this fraction is expressed as irreducible.

Finally, the sequence of transformations

$$(1) \qquad \begin{aligned} y &= x^{h} z_1 \\ x &= z_1^{h_1} z_2 \\ &\ldots\ldots\ldots \\ z_{s_1-1} &= z_{s_1}^{k_1} v \end{aligned}$$

leads us to the series $f_1'(u,v)$, where $u = z_{s_1}$.

Intuitively, $f_1'(u,v)$ may be thought therefore as an equation with centre $(0, a_{s_1 k_1})$ (see 1.3.9.) for the successive quadratic transform $\square_1'$ of $\square$ at the first leading free infinitely near point O_1' of the origin of the embedded curve $f=0$.

<u>Lemma 3.4.10.</u>- The composition of the transformations (1) is the new transformation given by:

$$y = u^{[h, h_1, \ldots, k_1]} \, v^{[h, h_1, \ldots, h_{s_1-1}]}$$

$$x = u^{[h_1, \ldots, k_1]} \, v^{[h_1, \ldots, h_{s_1-1}]}$$

<u>Proof:</u> By induction on the number N of transformations which are composed. For $N=1$, the result is evident. If it is true for $N=i$, we have

$$y = z_i^{[h,\ldots,h_i]} \, z_{i+1}^{[h,\ldots,h_{i-1}]}$$

$$x = z_i^{[h_1,\ldots,h_i]} \, z_{i+1}^{[h_1,\ldots,h_{i-1}]}$$

As $z_i = z_{i+1}^{h_{i+1}} \, z_{i+2}$, by making use of properties of the Euler polynomials, we obtain:

$$y = z_{i+1}^{[h,\ldots,h_{i+1}]} \, z_{i+2}^{[h,\ldots,h_i]}$$

$$x = z_{i+1}^{[h_1,\ldots,h_{i+1}]} \, z_{i+2}^{[h_1,\ldots,h_i]} .$$

Hence it is also true for $N = i+1$, and this completes the proof.

<u>Proposition 3.4.11.</u>- With notations as above the following statements hold:

(a) If $e_1 = (m,n)$ $(e_1 = n_{s_1})$, $m' = m/e_1$, $n' = n/e_1$, then $[h,h_1,\ldots,k_1] = m'$ and $[h_1,\ldots,k_1] = n'$.

(b) If $\sigma = [h,h_1,\ldots,h_{s_1-1}]$ and $\tau = [h_1,\ldots,h_{s_1-1}]$, then (σ,τ) is the unique solution of the diophantine equation

$$|\tau m' - \sigma n'| = 1$$

verifying the following two conditions:

(i) $\sigma \geqslant 0$, $\tau \geqslant 0$.

(ii) $2\tau \leqslant n'$, $2\sigma \leqslant m'$.

<u>Proof:</u> (a) is evident. We shall prove (b). Since

$$\frac{\sigma}{\tau} = \frac{[h,h_1,\ldots,h_{s_1-1}]}{[h_1,\ldots,h_{s_1-1}]}$$

it follows that σ/τ is the (s_1-1)-th aproximant fraction to m'/n'. (Wall, [23], page 15). Thus, using the wellknown properties of continued fractions we have

$$\frac{m'}{n'} - \frac{\sigma}{\tau} = \frac{(-1)^{s_1-1}}{n'}$$

or equivalently (σ, τ) is a solution of the diophantine equation

$$(2) \qquad \tau m' - \sigma n' = (-1)^{s_1-1} .$$

If (σ^*, τ^*) is another solution of (2), then

$$(\tau - \tau^*) m' = (\sigma - \sigma^*) n'$$

and hence $\sigma - \sigma^*$ and $\tau - \tau^*$ have the same sign. Furthermore, as $(m', n') = 1$, there exists $q \in \mathbf{Z}$ such that

$$(3) \qquad \begin{aligned} \sigma &= \sigma^* + q m' \\ \tau &= \tau^* + q n' \end{aligned}$$

i.e., with the product of the usual order on $\mathbf{Z} \times \mathbf{Z}$, we have

$$(\sigma^*, \tau^*) \leqslant (\sigma, \tau) \quad \text{or} \quad (\sigma^*, \tau^*) > (\sigma, \tau)$$

according as $q \geqslant 0$ or $q < 0$.

Relation (3) shows that (2) has only one solution verifying

$$(4) \qquad (0,0) \leqslant (\sigma, \tau) \leqslant (m', n').$$

Now, recall that our real solution is $\sigma = [h, \ldots, h_{s_1-1}]$, $\tau = [h_1, \ldots, h_{s_1-1}]$. It verifies the following inequalities:

$$m' = [h, h_1, \ldots, k_1] = k_1 [h, \ldots, h_{s_1-1}] + [h, \ldots, h_{s_1-2}] \geqslant 2\sigma .$$

$$n' = [h_1, \ldots, k_1] = k_1 [h_1, \ldots, h_{s_1-1}] + [h_1, \ldots, h_{s_1-2}] \geq 2\tau .$$

Finally, if we take a solution (σ_o, τ_o) (resp. (σ_1, τ_1)) of $\sigma n' - \tau m' = 1$ (resp. $\sigma n' - \tau m' = -1$) satisfying (4), then

$$\sigma_o + \sigma_1 = m'$$

$$\tau_o + \tau_1 = n'.$$

Therefore, only one of them can verify $(0,0) \leq 2(\sigma, \tau) \leq (m', n')$. That will be the right solution, and this completes the proof.

<u>Proposition 3.4.12.</u>- The plane linear transformation

$$T : (\alpha, \beta) \longrightarrow (\alpha n' + \beta m', \alpha\tau + \beta\sigma)$$

satisfies the following properties:

(a) The Newton polygon $\overline{AB}$ of f is transformed by T onto the vertical straight line segment $\overline{A'B'}$ joining $A' = (mn', m\tau)$ and $B' = (m'n, n\sigma)$.

The image of the Newton diagram of f is therefore contained in the external region bounded by the straight lines $\overline{OA'}$, $\overline{OB'}$ and $\overline{A'B'}$.

(b) If

$$f = \sum_{(\alpha,\beta)} A_{(\alpha,\beta)} X^{\alpha} Y^{\beta} ,$$

then,

$$f'_1 = \sum_{(\alpha',\beta')} A_{T^{-1}(\alpha',\beta')} U^{\alpha'-\delta} V^{\beta'-\gamma}$$

where $\delta = m.n/e_1$ and $\gamma = \min(\tau m, \sigma n)$.

(c) If

$$L(u,v,f'_1) = \sum_{(\alpha',\beta') \in \overline{A'B'}} A_{T^{-1}(\alpha',\beta')} u^{\alpha'-\delta} v^{\beta'-\gamma} ,$$

there exist $a, \lambda \in k$, $a \neq 0$, $\lambda \neq 0$, such that

$$L(u,v,f_1') = a\,(v-\lambda)^{e_1} .$$

<u>Proof:</u> Since $\begin{vmatrix} n' & m' \\ \tau & \sigma \end{vmatrix} = \pm 1$, T is bijective.

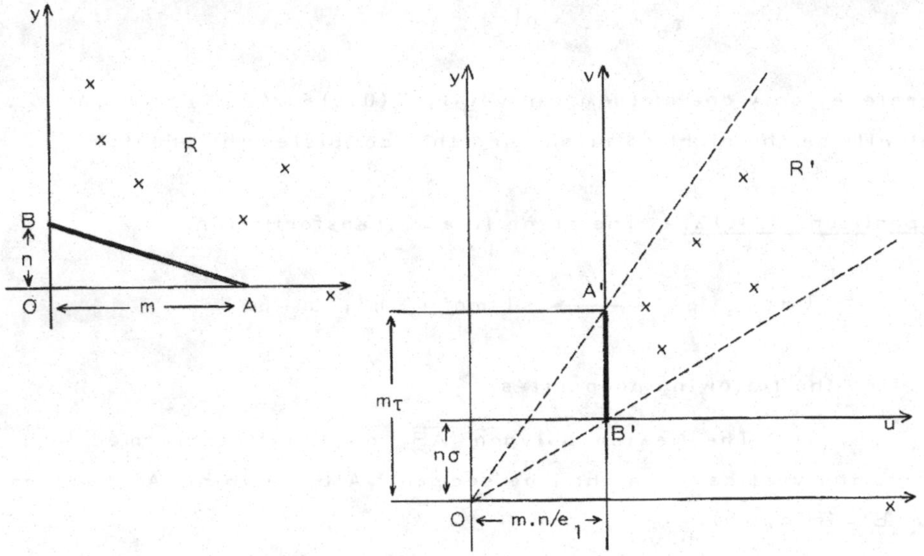

(a) It is trivial. (Notice that the length of $\overline{A'B'}$ is $e_1 = |\tau m - \sigma n|$).

(b) If $y = u^{m'} v^{\sigma}$, $x = u^{n'} v^{\tau}$, then

$$x^{\alpha} y^{\beta} = u^{n'\alpha + m'\beta} v^{\tau\alpha + \sigma\beta} .$$

It follows that the composition of quadratic transformations from $\{x,y\}$ to $\{u,v\}$ induces the transformation T on the plane. But, since $\{x,y\} \longrightarrow \{u,v\}$ transform f into f_1', the result is trivial.

(c) By 3.4.5. $L(x,y,f) = a\,(y^{n'} - \lambda x^{m'})^{e_1}$, with $a \neq 0$, and $\lambda \neq 0$. Hence,

$$L(x,y,f) = a\, u^{\delta}(v^{n'\sigma} - \lambda v^{m'\tau})^{e_1}.$$

If $\gamma = n\sigma$, then $L(u,v,f'_1) = a\,(1-\lambda v)^{e_1}$.

If $\gamma = m\tau$, then $L(u,v,f'_1) = a\,(v-\lambda)^{e_1}$.

In both cases (c) holds.

Remark 3.4.13.- The series f'_1 in (b) is a unit in $k((U,V))$. If $\lambda \in k$ is as in (c), setting $\bar{v} = v-\lambda$, then $\{u,\bar{v}\}$ is a basis of $\underline{m}'_1$, the maximal ideal of $\square'_1$.

The change $\bar{v} = v-\lambda$ acts on $D(f'_1)$ (Newton diagram of f'_1) exactly as the change 3.4.6. acts on $D(f)$.

Moreover, note that the substitution $\bar{v} = v - \lambda$ has a sense because $D(f'_1) \subseteq R'$ (see the picture), and thus

$$f'_1(u,v) = \sum_{j=0}^{\infty} p_j(v)\, u^j ,$$

where $p_j(v)$ is a polynomial in v for each j.

Algorithm 3.4.14.- Consider the curve $\square$ defined by the irreducible series $f(X,Y) \in k((X,Y))$, and keep the notations used from the beginning of this section.

Let $D(f)$ be the Newton diagram of f, and make successive substitutions of type 3.4.6. whenever n divides m. By 3.4.7., either:

(A) The process is infinite, or

(B) The process in finite, i.e., after a finite number of steps one obtains a diagram with a Newton polygon whose projections m_0 and $n=e_0$ on the coordinate axes verify $e_0 < m_0$ and $(e_0,m_0) < e_0$.

In the case (A) the algorithm finishes (this case holds if and only if $n=1$).

In the case (B), set $e_1 = (e_0, m_0)$, $m' = m_0/e_1$, $n' = e_0/e_1$, and solve the diophantine equation $|\tau m' - \sigma n'| = 1$ with the conditions 3.4.11. The linear transformation in 3.4.12. leads to a vertical polygon.

Make afterwards a new cycle of substitutions 3.4.6., being $\bar{v} = v - \lambda$, where λ is as in 3.4.12. (c), the first of these substitutions (see remark 3.4.13.).

Then, again either

(A_1) The process is infinite, or

(B_1) The process is finite, i.e., after a finite number of steps one obtains a diagram with a Newton polygon whose projections m_1 and e_1 on the axes verify $(e_1, m_1) < e_1$. (Note that it is not necessary that $m_1 > e_1$).

In the case (A_1) the algorithm finishes ($e_1 = 1$).

In the case (B_1), setting $e_2 = (e_1, m_1)$, it continues in the same way.

Since $e_0 > e_1 > \dots$, there exists an integer g' for which $e_{g'} = 1$. Thus the algorithm finishes after finitely many steps.

As a result of this algorithm g' polygons (and actually straight line segments) $(P_\nu)_{0 \leq \nu \leq g}$ are obtained. The lengths of the projections on the coordinate axes of these polygons are respectively e_ν and m_ν. Finally recall that:

(i) $(e_\nu, m_\nu) < e_\nu$, $0 \leq \nu \leq g'-1$.

(ii) $e_{\nu+1} = (e_\nu, m_\nu)$, $0 \leq \nu \leq g'-1$.

Theorem 3.4.15.- Let $\square$ be the algebroid curve defined by the irreducible series $f(X,Y) \in k((X,Y))$, $(\beta_\nu)_{0 \leq \nu \leq g}$ its characteristic exponents, and $(P_\nu)_{0 \leq \nu \leq g'-1}$ the polygons given by the above algorithm. Then:

(a) $g' = g$.

(b) $\beta_0 = e_0$, $\beta_{\nu+1} = m_0 + m_1 + \ldots + m_\nu$, $0 \leqslant \nu \leqslant g-1$.

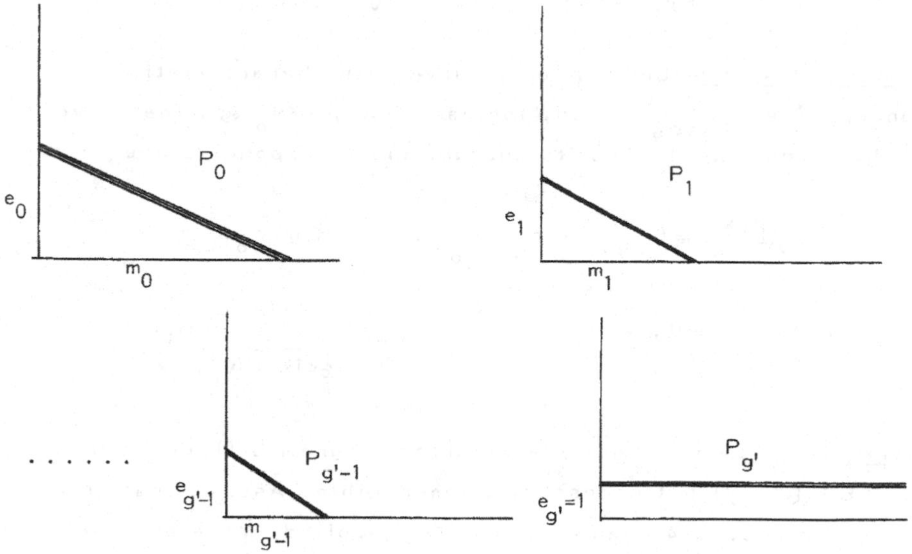

Proof: It is evident that $\beta_0 = e_0$. By 3.4.7. we have $\beta_1 = m_0$. Assume that the Hamburger-Noether expansion for $\square$ in the basis $\{x, y\}$ is written in the reduced form 3.3.4. We claim that $m_1 = (h_{s_1} - k_1)\, n_{s_1} + n_{s_1+1}$. Indeed the first characteristic exponent of $\square_1^*$ is given by

$$\beta_1(\square_1^*) = (h_{s_1} - k_1) n_{s_1} + n_{s_1+1} + n_{s_1} .$$

On the other hand, by construction of the polygon P_1, this exponent is also $m_1 + n_{s_1}$.

Now, if in our assumptions $\square$ is replaced by $\square_1^*$, we get $m_2 = (h_{s_2} - k_2)\, n_{s_2} + n_{s_2+1}$, and by an inductive method $m_\nu = (h_{s_\nu} - k_\nu)\, n_{s_\nu} + n_{s_\nu+1}$, for all ν, $0 \leqslant \nu \leqslant g'-1$. In particular $g = g'$.

Finally the formulae in proposition 3.3.6. lead us to the

expressions:

$$\beta_{\nu+1} = m_0 + m_1 + \dots + m_\nu \ , \quad 0 \leq \nu < g.$$

Corollary 3.4.16.- Let $\square$ be a curve with characteristic exponents $(\beta_\nu)_{0\leq\nu\leq g}$. Then, for each ν_o, $1 \leq \nu_o \leq g$, the curve $\square^*_{\nu_o}$ has genus $g-\nu_o$ and its characteristic exponents are

$$\beta_\nu(\square^*_{\nu_o}) = \beta_{\nu+\nu_o} - \beta_{\nu_o} + e_{\nu_o} \ , \quad 0 \leq \nu \leq g-\nu_o .$$

Proof: It is evident, since $\beta_{\nu+\nu_o} - \beta_{\nu_o} = \sum_{\nu_o \leq i < \nu+\nu_o} m_i$.

Remark 3.4.17.- Consider the weighted diagram D(f) of a series $f(X,Y) \in k((X,Y))$ not necessarily irreducible. Assume that on D(f) the algorithm 3.4.14. can be completely applied. More precisely, the Newton polygon is a straight line segment and the successive steps lead without any modifications to $e_{g'} = 1$. (Hence, the Newton polygon at each step is a straight line segment.)

Then, a Hamburger-Noether expansion may be associated to f in the following way:

By a transformation of type $y^* = y - a_{01}x - \dots - a_{0h}x^h$ we obtain a polygon with projections m and n, with $m > n$, and $(m,n) < n$. The top row of this Hamburger-Noether expansion is

$$y = a_{01}x + \dots + a_{0h}x^h + x^h z_1 .$$

Now, the representation of the rational number m/n by a continued fraction determines the following rows. Namely, if $h, h_1, \dots, h_{s_1-1}$, k_1 are the partial quotients, these rows are

$$z_{j-1} = z_j^{h_j} z_{j+1} \ , \quad 1 \leq j \leq s_1 - 1 .$$

The second step of the algorithm provides a change of type

$$v^* = v - a_{s_1 k_1} u - \ldots - a_{s_1, k_1+s_1} u^{l_1}$$

which yields the polygon P_1. We add the row

$$z_{s_1-1} = a_{s_1 k_1} z_{s_1}^{k_1} + \ldots + a_{s_1 h_{s_1}} z_{s_1}^{h_{s_1}} + z_{s_1}^{h_{s_1}} z_{s_1+1} ,$$

with $h_{s_1} = k_1 + l_1$.

The rest of the expansion is constructed in the same way. Note that since in the last step $e_g = 1$, the bottom row is of type

$$z_{s_g-1} = a_{s_g k_g} z_{s_g}^{k_g} + \ldots .$$

<u>Theorem 3.4.18.</u>- Keeping the assumptions as in the above remark, if

$$(\underline{D}) \qquad z_{j-1} = \sum_i a_{ji} z_j^i + z_j^{h_j} z_{j+1} , \quad 0 \leqslant j \leqslant r,$$

is the Hamburger-Noether expansion associated to f, then:

(a) f is irreducible.

(b) $(\underline{D})$ is the Hamburger-Noether expansion of the curve $\square = k((X,Y))/(f)$ in the basis $\{X + (f), Y + (f)\}$ of its maximal ideal.

<u>Proof:</u> $(\underline{D})$ defines a primitive parametric representation of an irreducible curve:

$$(1) \qquad \begin{aligned} x &= x(z_r) \\ y &= y(z_r) . \end{aligned}$$

The multiplicity of this curve is $n = \underline{u}(f)$. Hence, to prove that f is irreducible it suffices to check that $f(x(z_r), y(z_r)) = 0$, because this implies that f=0 is an equation for the irreducible curve (1).

Furthermore, the part (b) of the theorem also proceeds trivially from this equality.

We shall prove $f(x(z_r), y(z_r)) = 0$ by induction on n.

If $n = 1$, the formal change $\bar{Y} = Y - \sum_{i=0}^{\infty} a_{0i} X^i$ in $k[[X,Y]]$ transforms $f(X,Y)$ into a series $\bar{f}(X,\bar{Y})$ whose Newton polygon has no point on the x-axis. It follows that $\bar{f}(X,\bar{Y}) = \bar{Y}\, h(X,\bar{Y})$, and so,

$$f(X,Y) = \left(Y - \sum_{i=0}^{\infty} a_{0i} X^i \right) h'(X,Y) .$$

As the parametric equations (1) are actually

$$x = x$$
$$y = \sum_{i=0}^{\infty} a_{0i} x^i ,$$

then $f(x, y(x)) = 0$.

Now, let f be a series of order $n > 1$ satisfying the hypothesis of the theorem, and assume that the result is true for any g with $\underline{\upsilon}(g) < n$ satisfying also those hypothesis.

The changes $Y = Y - \sum_{i=0}^{h} a_{0i} X^i$, and $Y = U^{m'} V^{\sigma}$, $X = U^{n'} V^{\tau}$, where m', n', σ and τ are defined by the algorithm, transform $f(X,Y)$ into $U^{\delta} V^{\gamma} f_1'(U,V)$ (see 3.4.12.). By $\bar{V} = V - a_{s_1 k_1}$, $f_1'(U,V)$ turns into $f_1(U,\bar{V}) \in k[[U,\bar{V}]]$, being $\underline{\upsilon}(f_1) = e_1 = (m,n) < n$.

The series $f_1(U,\bar{V})$ allows the algorithm to continue, thus if we cut the expansion ($\underline{D}$) out through the corresponding term, the parametric equations

$$u = u(z_r)$$
$$\bar{v} = \bar{v}(z_r)$$

verify $f_1(u(z_r), \bar{v}(z_r)) = 0$ (by the induction hypothesis). It follows that $f(x(z_r),y(z_r)) = 0$ which completes the proof.

Corollary 3.4.19.- Let $f = f_1 + f_2 + \ldots \in k((X,Y))$ be a formal power series, where f_s is a homogeneous polynomial of degree s, $0 < s < \infty$. For each $m \geq 1$, set

$$h_m = f_1 + f_2 + \ldots + f_m.$$

Then, f is irreducible if and only if there exists an integer m_o such that h_m is irreducible for $m \geq m_o$.

Proof: The result derives directly from the above theorem since the property "the algorithm 3.4.14. can be applied to f" depends only on a finite set of points in $D(f)$.

Remark 3.4.20.- Theorem 3.4.18. points out how the algorithm 3.4.14. gives us the Hamburger-Noether expansion from the Newton diagram. In particular this algorithm can be viewed as a practical method to find parametric equations.

We also remark that if $f(X,Y) = 0$ is the equation of a reducible curve, the Hamburger-Noether expansions of the irreducible components of that curve can be obtained, in a similar fashion, from the Newton diagram for f.

5. CHARACTERISTIC EXPONENTS AND PUISEUX SERIES.

We have seen in chapter II that for a given curve it is not always possible to find a Puiseux series which represents it. In this fifth section we consider the curves for which Puiseux series exist and its characteristic exponents can be computed using these series.

Since in the characteristic zero case the above discussion is obvious, we shall assume in this section that k is an algebraically closed field of characteristic $p > 0$. The notations will be kept as in the preceding sections.

Theorem 3.5.1.- (PUISEUX'S THEOREM).- Let $\square$ be an irreducible plane algebroid curve over k. Assume that the multiplicity n of $\square$ is prime to p, and that

$$x = t^n$$
$$y = \sum_{i=n}^{\infty} a_i t^i$$

is a (primitive) parametric representation of $\square$ in a basis $\{x,y\}$ of its maximal ideal. Then, the characteristic exponents of the curve agree with the characteristic exponents of the above representation. (These have a sense according to 2.1.13. and 3.1.5.)

Proof: We shall use induction on n. For $n = 1$ the result is trivial since $\square$ is regular.

Let $\square$ be a curve of multiplicity $n > 1$, $(n,p) = 1$, which in the basis $\{x,y\}$ has a parametric representation

$$(1) \qquad x = t^n, \qquad y = \sum_{i=0}^{\infty} a_i t^i \, .$$

Assume that the theorem holds for any curve whose multiplicity is less than n and prime to p.

Denote by $(\beta_\nu)_{0 \leqslant \nu \leqslant g}$ (resp. $(\beta'_\nu)_{0 \leqslant \nu \leqslant g'}$) the characteristic exponents of $\square$ (resp. of (1)). We must prove that both sets of integers are the same. It is evident that $\beta_0 = \beta'_0 = n$, and $\beta_1 = \beta'_1 = m$.

We may assume, without loss of generality, that $a_i = 0$ for $i < m$, since these terms do not affect the values of β_ν nor those of β'_ν.

The proof will be long, so we shall divide it in several steps.

1st step. Outline.

Let $\square_1^*$ be the curve associated to the first terminal satellite infinitely near point (see 3.3.5.). The multiplicity of $\square_1^*$ is $n_{s_1} = e_1$, being $e_1 < n$, $(e_1, p) = 1$. Thus the induction hypothesis can be applied to $\square_1^*$.

A basis of the maximal ideal of $\square_1^*$ is $\{u, u.v\}$, where u and v are given by

$$\begin{aligned} y &= u^{m'} v^{\sigma} \\ x &= u^{n'} v^{\tau} \end{aligned} \quad ; \quad \tau m' - \sigma n' = \pm 1 \; ; \quad \begin{aligned} & m' = m/e_1 \; , \; n' = n/e_1 \; , \\ & 0 \leq 2\sigma \leq m' \; , \; 0 \leq 2\tau \leq n' . \end{aligned}$$

For an appropriate uniformizing $t' \in \overline{\square}$ we have

$$(2) \qquad \begin{aligned} u &= t'^{e_1} \\ v &= \sum_{i=0}^{\infty} b_i t'^i \; , \text{ with } b_0 \neq 0. \end{aligned}$$

According to 3.4.16. the curve $\square_1^*$ has genus $g-1$ and its characteristic exponents $(\beta_\nu^*)_{0 \leq \nu \leq g-1}$ are given by

$$\beta_\nu^* = (\beta_{\nu+1} - \beta_1) + e_1 .$$

By the induction hypothesis these are the characteristic exponents of the representation

$$\begin{aligned} u &= t'^{e_1} \\ u.v &= \sum_{i=0}^{\infty} b_i t'^{i+e_1} . \end{aligned}$$

2nd step. Auxiliar curves $\widehat{\square}$ and $\widehat{\square}'$.

Since t and t' are both uniformizing parameters for $\square$, then t/t' and t'/t are units in $\overline{\square}$.

Consider the auxiliar curves $\widehat{\square}$ and $\widehat{\square}'$ defined respectively by

$$(3)\qquad \widehat{\square}: \begin{cases} \hat{u} = t^{e_1} \\ \hat{v} = (t'/t)\, t^{e_1} \end{cases} \qquad\qquad \widehat{\square}': \begin{cases} \hat{u}' = t'^{e_1} \\ \hat{v}' = (t/t')\, t'^{e_1} \end{cases}$$

These representations are primitive (this fact is not trivial and it will be proved later). We shall see now that $\widehat{\square}$ and $\widehat{\square}'$ are (a)-equisingular. Indeed, since they have the same multiplicity, it suffices to prove that its respective quadratic transforms are (a)-equisingular. But this is evident, because

$$\widehat{\square}_1 = k\left(\left(t^{e_1}, \frac{t'}{t} - \frac{\hat{v}(0)}{\hat{v}'(0)}\right)\right) = k\left(\left(t'^{e_1}, \frac{t}{t'} - \frac{\hat{v}'(0)}{\hat{v}(0)}\right)\right) = \widehat{\square}'_1 .$$

Hence the characteristic exponents of the two representations (3) agree, since by the induction hypothesis they are respectively the characteristic exponents of the two (a)-equisingular curves $\widehat{\square}$ and $\widehat{\square}'$.

3^{rt} step. Relation between $\square^*_1$ and $\widehat{\square}'$.

For each ν, $1 \leq \nu \leq g-1$, let $e_\nu = (\beta^*_0, \dots, \beta^*_{\nu-1}) = (\beta_0, \dots, \beta_\nu)$, and let w_ν be a primitive e_ν-th root of unity (note that $(e_\nu, p) = 1$). Let f_ν be the k-automorphism of $k[[t']]$ defined by $f_\nu(t') = w_\nu t'$.

Since $v^{\tau} = x/_{u} n'\sigma$, we have $v^{\tau} = (t/t')^n$. Then:

$$(4)\qquad f_\nu(v)^\tau - v^\tau = f_\nu(t/t')^n - (t/t')^n .$$

If $\tau = \bar{\tau} p^r$ with $(\bar{\tau}, p) = 1$, and $\underline{v}$ is the natural valuation associated with $\square$, the value in $\underline{v}$ of the left hand side member of (4) is

$$p^r\,\underline{v}(f_\nu(v)^{\overline{\tau}} - v^{\overline{\tau}}) = p^r\,\underline{v}(f_\nu(v) - v) = p^r\,(\beta_\nu^* - e_1).$$

(This equality is verified because

$$f_\nu(v)^{\overline{\tau}} - v^{\overline{\tau}} = (f_\nu(v) - v).v^*$$

where $v^* = \sum_{i+j=\overline{\tau}-1} v^i\,(f_\nu(v))^j$, with $v^*(0) = \overline{\tau}\,v(0) \neq 0$).

The $\underline{v}$-value of the right hand side member of (4) is, by the same reason,

$$\underline{v}(\,f_\nu(t/t') - (t/t'))\,.$$

Now, the equalities $p^r\,(\beta_\nu^* - e_1) = \underline{v}(\,f_\nu(t/t') - (t/t'))$, $1 \leqslant \nu \leqslant g-1$, imply that the representation (3) for $\widehat{\square}'$ has the following system of characteristic exponents:

$$\{\,p^r(\,\beta_\nu^* - e_1) + e_1\}_{0 \leqslant \nu \leqslant g-1}\,.$$

In particular, by using 2.1.13. this representation is primitive, and since $\widehat{\square}_1 = \widehat{\square}'_1$, so is the other representation (3) . Recall that the proof of these points had been left out in the second step.

4th step. Relation between $\square$ and $\widehat{\square}$.

Let g_ν be the k-automorphism of $k((t))$ defined by $g_\nu(t) = w_\nu\,t$, $1 \leqslant \nu \leqslant g-1$.

As $y = u^{m'}v^{\sigma}$, we have $y = u^{m'\tau}v^{n'\sigma} = t'^{m\tau}(t/t')^n$. Then,

$$(y/_t m)^\tau = (t'/t)^{\varepsilon e_1}\,,$$

where $\varepsilon = \pm 1$ according as $\tau m - \sigma n = \pm\, e_1$. It follows that

$$(5) \qquad g_\nu(y/_t m)^\tau - (y/_t m)^\tau = g\,(t'/t)^{\varepsilon e_1} - (t'/t)^{\varepsilon e_1}\,.$$

The $\underline{v}$-value of the left hand side member of (5) is

$$p^r\, \underline{v}(g_\nu(y/_t m) - (y/_t m)) = p^r\,(\beta'_{\nu+1} - m) = p^r(\beta'_{\nu+1} - \beta'_1).$$

The $\underline{v}$-value of the right hand side member does not depend on ε, since t'/t and $g_\nu(t'/t)$ are units in $\overline{\square}$, and as $(e_1,p)=1$, it is

$$\underline{v}(\,g_\nu(t'/t)-(t'/t)).$$

As above the equalities

$$p^r\,(\,\beta'_{\nu+1} - \beta'_1) = \underline{v}(\,g_\nu(t'/t)-(t'/t))\quad,\quad 1 \leqslant \nu \leqslant g-1,$$

imply that the system of characteristic exponents of $\hat{\square}$ is

$$\{\,p^r\,(\,\beta'_{\nu+1} - \beta'_1) + e_1\}_{0 \leqslant \nu \leqslant g'-1}\,.$$

5th step. Conclusion.

The characteristic exponents of $\hat{\square}$ and $\hat{\square}'$ are the same (step 2). Thus, by using the formulae in the third and fourth steps we have $g' = g$ and

$$p^r\,(\,\beta^*_\nu - e_1) + e_1 = p^r\,(\,\beta'_{\nu+1} - \beta'_1) + e_1\,,\quad 1 \leqslant \nu \leqslant g-1,$$

whence

$$\beta^*_\nu = \beta'_{\nu+1} - \beta'_1 + e_1\quad,\quad 1 \leqslant \nu \leqslant g-1.$$

But, in the step 1 we have pointed out that $\beta_1 = \beta'_1$ and that

$$\beta^*_\nu = \beta_{\nu+1} - \beta_1 + e_1\,,$$

then $\beta_{\nu+1} = \beta'_{\nu+1}$, $1 \leqslant \nu \leqslant g-1$. This completes the proof of the theorem.

Corollary 3.5.2.- Let $\square$ be a curve, which has in the basis $\{x,y\}$ of its maximal ideal a parametric representation of type

$$x = \sum_{i=n}^{\infty} b_i t^i , \quad b_n \neq 0,$$
$$y = t^m$$

where $n < m$ and $(m,p) = 1$.

If $(\beta_\nu')_{0 \leq \nu \leq g}$ $(\beta_0' = m)$ are the characteristic exponents of this representation and $(\beta_\nu)_{0 \leq \nu \leq g}$ $(\beta_0 = n)$ the ones of $\square$, then:

(a) If n divides m,

$g = g'-1$, $\beta_0 = \beta_1'$ and $\beta_\nu = \beta_{\nu+1}' + m - n$, $1 \leq \nu \leq g-1$,

(b) If n does not divide m,

$g = g'$, $\beta_0 = \beta_1'$ and $\beta_\nu = \beta_\nu' + m - n$, $1 \leq \nu \leq g$.

Proof: Let $\square'$ be the curve defined by

$$x = \sum_{i=n}^{\infty} b_i t^{i+m}$$
$$y = t^m .$$

According to the above theorem, the characteristic exponents of $\square'$ are $\beta_0^* = m$ and $\beta_\nu^* = \beta_\nu' + m$, $1 \leq \nu \leq g'$. Now, since $\square_1' = \square$, the proof follows from lemma 3.2.11.

Remark 3.5.3.- The corollary shows how, by using Puiseux series, the characteristic exponents may be computed also in the case $(\beta_1, p) = 1$ although $(n,p) > 1$.

Thus, this corollary may be considered as a "generalized inversion formula" for Puiseux series.

The following examples show that if $(n,p) > 1$ and Puiseux expansions exist, the characteristic exponents of these expansions have no relation with the characteristic exponents of the curve.

Example 3.5.4.- Let k be an algebraically closed field of characteristic $p > 0$, and consider the curve $\square$ over k given by

$$(1) \qquad \begin{aligned} x &= t^{p^3} \\ y &= t^{p^3+p^2} + t^{p^3+p^2+p+1} \end{aligned}$$

First, since this parametric representation is primitive (see 2.1.13.), the multiplicity of $\square$ is p^3. Its Hamburger-Noether expansion is

$$\begin{aligned} y &= x z_1 \\ x &= z_1^p + z_1^{p+1} z_2 \\ z_1 &= -z_2^p + z_2^{p+1} z_3 \qquad\qquad (\underline{v}(x) = p^3, \underline{v}(z_1) = p^2, \underline{v}(z_2) = p, \underline{v}(z_3) = 1). \\ z_2 &= z_3^p + \ldots . \end{aligned}$$

Thus, the curve has genus 3 and its characteristic exponents are:

$$\begin{aligned} \beta_0 &= p^3 \\ \beta_1 &= p^3 + p^2 \\ \beta_2 &= p^3 + 2p^2 + p \\ \beta_3 &= p^3 + 2p^2 + 2p + 1 , \end{aligned}$$

while the characteristic exponents of the Puiseux parametric representation (1) are

$$\begin{aligned} \beta'_0 &= p^3 \\ \beta'_1 &= p^3 + p^2 \\ \beta'_2 &= p^3 + p^2 + p + 1 . \end{aligned}$$

Now, we summarize the anomalies of the curve $\square$:

(i) It has genus 3, while the Puiseux series has only two characteristic exponents.

(ii) β_2 and β_3 do not occur as effective exponents in the Puiseux series.

(iii) For no parameter $z \in \underline{m}$ ($\underline{m}$ being the maximal ideal of $\square$), neither β_2 nor β_3 occur as effective exponents in the expression of z as series in t.

This follows from the fact that $z = at^{p^3} + bt^{p^3+p^2} + bt^{p^3+p^2+p+1} + g(t)$ where $a, b \in k$ and $\underline{\upsilon}(g) > \beta_3$ if $p \geqslant 3$. (If $p=2$ (iii) is evident).

<u>Example 3.5.5.</u>- Let R and R' be two parametric representations of Puiseux type . Between the following statements there is no relation:

(a) R and R' have the same characteristic exponents (as Puiseux series).

(b) The algebroid curves defined respectively by R and R' are (a)-equisingular.

In fact, let k be a field of characteristic $p \geqslant 3$.

<u>(a) $\not\Rightarrow$ (b)</u> Consider the following parametric representations:

$$(R) \qquad \begin{aligned} x &= t^{p^3} \\ y &= t^{p^3+p^2} + t^{p^3+p^2+p} + t^{p^3+p^2+p+1} \end{aligned}$$

$$(R') \qquad \begin{aligned} x &= t^{p^3} \\ y &= t^{p^3+p^2} + t^{p^3+p^2+p} - t^{p^3+p^2+p+1} \end{aligned}$$

They evidently have the same set of characteristic exponents, but they are not (a)-equisingular. Indeed, the two top rows in the respective Hamburger-Noether expansions are both

$$y = x z_1$$
$$x = z_1^p - z_1^{p+1} + z_1^{p+1} z_2 ,$$

however, in the first case $\underline{v}(z_2) = p+1$ while in the second $\underline{v}(z_2)=p$.

<u>(b) $\not\Rightarrow$ (a)</u> Consider the representation R' as above, and the representation R'' used in 3.5.4. and again $p \geqslant 3$. The Hamburger-Noether expansions are

(R')
$$y = x z_1$$
$$x = z_1^p - z_1^{p+1} + z_1^{p+1} z_2$$
$$z_1 = (1/2) z_2^p + (3/8) z_2^{p+1} + z_2^{p+1} z_3$$
$$z_2 = -32 z_3^p + \dots .$$

(R'')
$$y = x z_1$$
$$x = z_1^p + z_1^{p+1} z_2$$
$$z_1 = -z_2^p + z_2^{p+1} z_3$$
$$z_2 = z_3^p + \dots .$$

R' and R'' do not have the same characteristic exponents but they define (a)-equisingular plane algebroid curves.

CHAPTER IV

OTHER SYSTEMS OF INVARIANTS FOR THE EQUISINGULARITY OF PLANE ALGEBROID CURVES.

In this chapter we consider new complete systems of invariants for the equiresolution of plane algebroid curves and their relationship with the characteristic exponents.

The first of them is explained in the first section and it is formed by the Newton coefficients given by Lejeune in [15]. In the second section we study briefly the maximal contact of any genus, using Hamburger-Noether expansions. The third section is devoted to the semigroup of values of a plane curve. We compute it and obtain results analogous to those of the characteristic zero case. Finally in the fourth section we generalize to the case of positive characteristic the main properties of the degree of the conductor of any curve.

1. NEWTON COEFFICIENTS.

As we have pointet out, this section is devoted to the study of Newton coefficients of an irreducible plane algebroid curve (Lejeune, [15]), and their expression in terms of the characteristic exponents.

Let C be an embedded irreducible plane algebroid curve defined over the algebraically closed field k by the series $f(X,Y) \in k[[X,Y]]$. Denote by $\square$ its local ring and set $x=X+(f)$, $y=Y+(f)$.

For any embedded curve Γ we shall denote by (Γ, C) the intersection multiplicity of Γ and C. By varying Γ to be regular, we define

$$\overline{\beta}_1(C) = \sup_{\Gamma \text{ reg.}} (C, \Gamma).$$

Definition 4.1.1.- A regular curve Γ is said to have the maximal contact with C iff $(C, \Gamma) = \overline{\beta}_1(C)$.

If C is regular, the only curve which has the maximal contact with C is itself, and actually $\overline{\beta}_1(C) = \infty$. If C is not regular, requirements for the maximal contact are given in the following:

Lemma 4.1.2.- Let Γ be a regular curve and assume that C is not regular. The following statements:

(i) The multiplicity of C does not divide (C, Γ),

(ii) Γ has the maximal contact with C,

are equivalent.

Proof: By using an appropriate formal change of variables in $k[[X, Y]]$ we may suppose, without loss of generality, that x is a transversal parameter for $\square$ and that y=0 is an equation for Γ. Now, the proof follows from 3.4.7.

Remark 4.1.3.- The results of the section 4 in the preceding chapter show also that if C is not regular, then:

$$\overline{\beta}_1(C) = hn + n_1 = \beta_1 \qquad , \qquad n = \underline{\nu}(f) = e(C).$$

Definition 4.1.4.- The first Newton coefficient for C is defined to be the rational number (and actually not integer) given by:

$$\mu_1(C) = \frac{\overline{\beta}_1(C)}{n} = \frac{hn + n_1}{n} .$$

We may write $\mu_1(□)$ since this value is intrinsic of the local ring □.

If Γ is a regular curve having the maximal contact with C, then $\mu_1(C) = \frac{(C,\Gamma)}{n}$. Moreover, by turning back to the Newton diagram, if x is transversal then $y=0$ has the maximal contact with C if and only if the Newton polygon of f joins the points $A=(m,0)$ and $B=(0,n)$ with $(m,n) < n$.

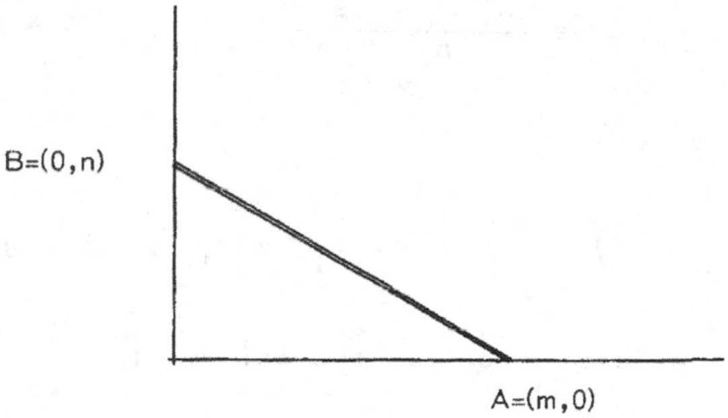

<u>Notations 4.1.5.</u>- Let us consider a plane curve □ and let

$$□ \subset □_1 \subset \ldots \subset □_M = \overline{□}$$

be its desingularization sequence 1.5.10.

According to the definition of genus and as under a quadratic transformation it decreses by at most a unit, there exist integers $0 = a_1 < a_2 < \ldots < a_g$, such that:

(i) $g = g(□)$.

(ii) $g(□_{a_\nu}) = g+1-\nu$, $1 \leqslant \nu \leqslant g$.

(iii) $g(□_j) = g+1-\nu$, if $a_\nu \leqslant j < a_{\nu+1}$.

<u>Definition 4.1.6.</u>- Let □ be a plane curve. Keeping notations as above, the <u>ν-th Newton coefficient of □</u>, $\mu_\nu(□)$, is defined to be the first Newton coefficient of the curve $□_{a_\nu}$.

Proposition 4.1.7.- Assume that in the basis $\{x,y\}$ of the maximal ideal $\underline{m}$ of $\square$, the curve $\square$ has a Hamburger-Noether expansion given by

$$(D') \qquad z_{j-1} = \sum_i a_{ji} z_j^i + z_j^{h_j} z_{j+1} \quad , \quad 0 \leqslant j \leqslant r.$$

Assume that (D') is written in the reduced form 3.3.4. Then,

$$\mu_{\nu+1}(\square) = \frac{h_{s_\nu} n_{s_\nu} + n_{s_\nu+1}}{n_{s_\nu}} \quad , \quad 1 \leqslant \nu \leqslant g-1.$$

Proof: It is evident, since

$$z_{j-1} = \sum_i a_{ji} z_j^i + z_j^{h_j} z_{j+1} \quad , \quad s_\nu \leqslant j \leqslant r,$$

is a Hamburger-Noether expansion for the curve $\square_{a_{\nu+1}}$.

Corollary 4.1.8.- Let $\square$ be a curve of genus g, $(\beta_\nu)_{0 \leqslant \nu \leqslant g}$ its characteristic exponents and $(\mu_\nu)_{1 \leqslant \nu \leqslant g}$ its Newton coefficients. Then,

$$\beta_1 = \beta_0 \cdot \mu_1 \quad ,$$
$$\beta_\nu = \beta_{\nu-1} + e_{\nu-1}(\mu_\nu - k_{\nu-1}).$$

Proof: It proceeds trivially form 3.3.7. and the above proposition.

Theorem 4.1.9.- Two given curves are (a)-equisingular if and only if they have the same genus and the same Newton coefficients.

Proof: Necessity is obvious from the proposition 4.1.7. Conversely, we claim that the Newton coefficients determine the characteristic exponents.

In fact, if $\mu_\nu = \frac{m'_\nu}{n'_\nu}$ with $(m'_\nu, n'_\nu) = 1$, we have $n'_\nu = e_{\nu-1}/e_\nu$, so

$$\beta_0 = n = n'_1 . n'_2 . \ldots . n'_g .$$

Now, the β_ν can be computed inductively from β_0 and the formulae in 4.1.8. (Recall that $\beta_0, \ldots, \beta_\nu$ determine $e_{\nu-1}$ and k_ν, (see 3.3.9.).)

Remark 4.1.10.- Formulae in 4.1.8. can be viewed as inversion formulae for characteristic exponents and Newton coefficients.

2. MAXIMAL CONTACT OF HIGHER GENUS.

Let us consider an embedded irreducible plane algebroid curve C over the algebraically closed field k. Denote by $\square$ its local ring.

Let g be the genus of $\square$ and γ an integer, $0 \leq \gamma \leq g$. If $(\beta_\nu)_{0 \leq \nu \leq g}$ are the characteristic exponents of $\square$, for each ν, $0 \leq \nu \leq \gamma$, we put

$$\beta^*_\nu = \frac{\beta_\nu}{(\beta_0, \ldots, \beta_\gamma)} = \frac{\beta_\nu}{e_\gamma} .$$

Denote by $\mathfrak{C}_\gamma(C)$ the set of irreducible curves Γ of genus γ embedded in the ambient plane having $(\beta^*_\nu)_{0 \leq \nu \leq \gamma}$ as system of characteristic exponents.

The set $\mathfrak{C}_\gamma(C)$ is non-empty. In fact, if $k = \mathbb{C}$, since

(i) $\beta^*_0 < \ldots . < \beta^*_\gamma$,

(ii) $\beta^*_0 > (\beta^*_0, \beta^*_1) > \ldots . > (\beta^*_0, \ldots, \beta^*_\gamma) = 1$,

the curve $x = t^{\beta^*_0}$, $y = \sum t^{\beta^*_i}$ is in $\mathfrak{C}_\gamma(C)$. If $k \neq \mathbb{C}$, by using an argument as in 3.2.4. one may see that this curve is a complex model for some curve over k, which is clearly in $\mathfrak{C}_\gamma(C)$.

Definition 4.2.1.- We define the value of the maximal contact of genus γ to be the number

$$\overline{\beta}_{\gamma+1} = \sup_{\Gamma \in \mathcal{E}_\gamma(C)} (C,\Gamma) .$$

Definition 4.2.2.- A curve $\Gamma \in \mathcal{E}_\gamma(C)$ is said to have the maximal contact of genus γ with C iff

$$(C,\Gamma) = \overline{\beta}_{\gamma+1}(C) .$$

Remark 4.2.3.- 1) If $\gamma = 0$, since $\mathcal{E}_0(C)$ is the set of regular curves, the value of the maximal contact of genus 0 is nothing but the number defined by 4.1.1. In the same way, the regular curves having the maximal contact with C are the curves in $\mathcal{E}_0(C)$ having the maximal contact of genus 0.

2) If $\gamma = g$, $\overline{\beta}_{g+1}(C) = \infty$, and C is itself the only curve which has the maximal contact of genus g.

Proposition 4.2.4.- Let Γ be a curve of genus γ. The following statements

(a) $\beta_\nu(\Gamma) = \beta^*_\nu$, $0 \leq \nu \leq \gamma$.

(b) $\mu_\nu(\Gamma) = \mu_\nu(C)$, $1 \leq \nu \leq \gamma$.

are equivalent. ($\mu_\nu(—)$ denotes the ν-th Newton coefficient).

Proof: Let

$$z_{j-1} = \sum_i a_{ji} z_j^i + z_j^{h_j} z_{j+1} , \quad 0 \leq j \leq r,$$

be the Hamburger-Noether expansion of C. We shall use the habitual notations.

Let C_γ be the curve whose Hamburger-Noether expansion is

$$\bar{z}_{j-1} = \sum_i a_{ji} \bar{z}_j^{\,i} + \bar{z}_j^{\,h_j} \bar{z}_{j+1} \quad , \quad 0 \leqslant j \leqslant s_\gamma - 1,$$

$$\bar{z}_{s_\gamma - 1} = a_{s_\gamma, k_\gamma} \bar{z}_{s_\gamma}^{\,k_\gamma} + \ldots + a_{s_\gamma, h_{s_\gamma}} \bar{z}_{s_\gamma}^{\,h_{s_\gamma}} \quad .$$

It is trivial that $C_\gamma \in \mathcal{E}_\gamma(C)$, since by 3.3.7. we have

$$\beta_\nu(C_\gamma) = \frac{\beta_\nu}{e_\gamma} = \beta^*_\nu \quad , \quad 0 \leqslant \nu \leqslant \gamma.$$

On the other hand, using 4.1.7. it is also tivial that

$$\mu_\nu(C_\gamma) = \mu_\nu(C) \quad , \quad 1 \leqslant \nu \leqslant \gamma .$$

Now, the proof follows inmediatly, because according to 3.2.10. and 4.1.9. each one of the statements (a),(b) is equivalent to:

" Γ and C_γ are (a)-equisingular".

<u>Remark 4.2.5.-</u> $\mathcal{E}_\gamma(C)$ may be also defined to be the set of irreducible curves of genus γ whose Newton coefficients agree with the γ first ones of C. Even, it is the set of curves which are (a)-equisingular to C_γ.

<u>Corollary 4.2.6.-</u> Let

$$\text{(D')} \qquad z_{j-1} = \sum_i a_{ji} z_j^{\,i} + z_j^{\,h_j} z_{j+1} \,, \quad 0 \leqslant j \leqslant r,$$

the Hamburger-Noether expansion of C, written in the reduced form 3.3.4. Then $\Gamma \in \mathcal{E}_\gamma(C)$ if and only if its Hamburger-Noether expansion is of type:

$$\text{(1)} \qquad \begin{aligned} \bar{y} &= b_{01} \bar{x} + \ldots + b_{0h} \bar{x}^{h} + \bar{x}^{h} \bar{z}_1 \\ \bar{x} &= \bar{z}_1^{\,h_1} \bar{z}_2 \\ &\ldots\ldots\ldots\ldots \\ \bar{z}_{s_1 - 1} &= b_{s_1 k_1} \bar{z}_{s_1}^{\,k_1} + \ldots + b_{s_1 h_{s_1}} \bar{z}_{s_1}^{\,h_{s_1}} + \bar{z}_{s_1}^{\,h_{s_1}} \bar{z}_{s_1+1} \end{aligned}$$

. . . .

$$\bar{z}_{s_\gamma - 1} = b_{s_\gamma k_\gamma} \bar{z}_{s_\gamma}^{k_\gamma} + \ldots .$$

with $b_{s_\nu k_\nu} \neq 0$, $1 \leq \nu \leq \gamma$.

Proposition 4.2.7.- Keeping the hypothesis and notations as above, if $0 \leq \gamma < g$, and

$$S_\gamma = \frac{1}{n_{s_\gamma}} \left(\sum_{j=0}^{s_\gamma} h_j n_j^2 \right) + n_{s_\gamma + 1} ,$$

then:

(a) If $\Gamma \in \xi_\gamma(C)$, $(\Gamma, C) \leq S_\gamma$.

(b) $(C_\gamma, C) = S_\gamma$.

Proof: Taking into account that $n_j(\Gamma) = n_j / n_{s_\gamma}$, $0 \leq j \leq s_\gamma$, (a) follows from the intersection multiplicity formula 2.3.3.

Notice that the equality holds if and only if in (1), $a_{ji} = b_{ji}$, for $j \leq s_\gamma$ and $i \leq h_j$. In particular (b) follows.

Corollary 4.2.8.- For each γ , $0 \leq \gamma < g$, we have

$$S_\gamma = \bar{\beta}_{\gamma+1}(C).$$

In particular, the values of the maximal contact are invariants of the local ring $\square$. We may write therefore $\bar{\beta}_{\gamma+1}(\square)$.

Theorem 4.2.9.- Let C be a curve with Hamburger-Noether expansion given by

$$z_{j-1} = \sum_i a_{ji} z_j^i + z_j^{h_j} z_{j+1} , \quad 0 \leq j \leq r.$$

Then, another curve $\Gamma \in \xi_\gamma(C)$ has the maximal contact of genus γ

with C if and only if its Hamburger-Noether expansion is of type

$$\bar{z}_{j-1} = \sum_i a_{ji}\, \bar{z}_j^{\,i} + \bar{z}_j^{\,h_j}\, \bar{z}_{j+1}\ , \quad 0 \leqslant j \leqslant s_\gamma - 1\ ,$$

$$\bar{z}_{s_\gamma - 1} = \sum_{1 \leqslant i \leqslant h_{s_\gamma}} a_{s_\gamma i}\, \bar{z}_{s_\gamma}^{\,i} + g(\bar{z}_{s_\gamma})\ ,$$

where g is a series in $\bar{z}_{s_\gamma}$ of order $> h_{s_\gamma}$.

Proof: It is trivial by using the remark made in the proof of proposition 4.2.7.

Remark 4.2.10.- From the above theorem we may deduce that a curve having the maximal contact with C of any genus exists, but it is not, of course, unique.

The maximal contact of higher genus may be sketched in the terminology of infinitely near points (see 1.5.15. and 1.5.16.):

"Curves having the maximal contact of genus γ with C are nothing but curves of genus γ for which the sequence of infinitely near points of its origin contains the $h+h_1+\ldots+h_{s_\gamma}$ first points of the sequence of infinitely near points of the origin of C (i.e., the points of this sequence with multiplicity $\geqslant e_\gamma = n_{s_\gamma}$)".

Remark 4.2.11.- Most of the main properties of the maximal contact shown by Lejeune in (15) can be obtained inmediately from theorem 4.2.9. We summarize some of these properties:

4.2.11.1. Let C_1 be a curve having maximal contact of genus $g_1 \leqslant g$ with C, then if a curve has the maximal contact of genus $\gamma \leqslant g_1$ with C_1, it also has it with C. In particular, for $\gamma = 0$, if C is refered to $\{x, y\}$ and if $y = 0$ has the maximal contact with C_1, so has it with C.

4.2.11.2. Let C_1 be a curve having maximal contact with C of genus $g_1 \leq g$. Then, any curve having the maximal contact with C of genus $\gamma < g_1$, so has it with C_1.

Note that if $\gamma = g_1$ the statement does not hold since curves having the maximal contact of genus g_1 with C are not unique in general.

<u>4.2.11.3.</u> The multiplicity of a curve Γ having the maximal contact of genus γ with C is n/e_γ.

<u>4.2.11.4.</u> Let C be a curve of genus g, and Γ another curve of genus γ, $1 \leq \gamma < g$. Consider a formal quadratic transformation in the ambient plane and denote by C_1 and Γ_1 their respective strict transforms. Then Γ has the maximal contact with C if and only if:

(i) $\mu_1(\Gamma) = \mu_1(C)$ (or $\beta_1(\Gamma) = \beta_1^*$).

(ii) Γ_1 has the maximal contact with C_1.

To show the sufficient condition note that, since Γ_1 and C_1 have the maximal contact, they have the same origin, and so $a_{01} = b_{01}$.

<u>4.2.11.5.</u> Assume that Γ has the maximal contact of genus γ with C. Let

$$(1) \quad \begin{array}{l} C_{M(C)} \longrightarrow C_{M(C)-1} \longrightarrow \dots \longrightarrow C_1 \longrightarrow C \\ \Gamma_{M(\Gamma)} \longrightarrow \Gamma_{M(\Gamma)-1} \longrightarrow \dots \longrightarrow \Gamma_1 \longrightarrow \Gamma \end{array}$$

be respective sequences of strict transforms by a sequence of formal quadratic transformations which is a sequence of desingularization (1.5.10.) simultaneosly for C and Γ. Then for each i, $0 \leq i \leq M(\Gamma)$, we have $g(\Gamma) = g(C_i) + \gamma - g$. (This follows from the definition of the genus and the fact that $e(\Gamma_i)/_{e(\Gamma)} = e(C_i)/_{e(C)}$, which is a trivial consequence of 4.2.9.)

4.2.11.6. Let Γ a curve with genus γ and C another one with genus $g, \gamma < g$. Assume that (1) are respective desingularization sequences. Furthermore, assume that the $M(\Gamma)$ first transformations of the sequence for C agree with those for Γ.

If $\Gamma \in \mathcal{E}_\gamma(C)$ and if $\Gamma_{M(\Gamma)}$ has the maximal contact (of genus 0) with $C_{M(\Gamma)}$ then Γ has the maximal contact (of genus γ) with C. (In fact, the condition implies that Γ and C have $h + h_1 + \ldots + h_{s_\gamma}$ infinitely near points in common, and hence the result follows from 4.2.10.)

4.2.11.7. Let C a curve with genus g, and

$$C_{M(C)} \longrightarrow C_{M(C)-1} \longrightarrow \ldots \longrightarrow C_1 \longrightarrow C$$

a sequence of desingularization for it. For each $\gamma < g$, set $a_\gamma = \inf \{ i \mid g(C_i) = g - \gamma \}$. If U is a regular curve having the maximal contact with C_{a_γ}, then there exists a curve Γ of genus γ having the maximal contact with C, such that if Γ_{a_γ} is the a_γ-th transform of Γ by the above sequence, then $\Gamma_{a_\gamma} = U$. (To prove this, note that by 4.2.9. the Hamburger-Noether expansion of C together with the Hamburger-Noether expansion of U determine that of Γ.)

3. THE SEMIGROUP OF VALUES.

Let us consider an irreducible plane algebroid curve $\square$ over the algebraically closed field k. Denote by $\overline{\square}$ the integral closure of $\square$ in its quotient field F, and by $\underline{v}$ the natural valuation of F:

$$\underline{v} : F \longrightarrow \mathbf{Z}$$

If t is any uniformizing parameter for $\square$, we have $\overline{\square} = k[[t]]$, $F = k((t))$ and $\underline{v}(z) = \underline{\upsilon}(z)$ for any series $z = z(t) \in \overline{\square}$.

Since $\square$ is an integral domain and $\square \subset \overline{\square} \subset F$ the set $\underline{v}(\square - \{0\})$ is an additive subsemigroup of $\mathbb{Z}_+$.

<u>Definition 4.3.1.-</u> The set $S(\square) = \underline{v}(\square - \{0\}) \subset \mathbb{Z}_+$ will be called <u>semigroup of values of</u> $\square$.

<u>Remarks and notations 4.3.2.-</u> Let C be an embedded irreducible plane algebroid curve defined by $f(X,Y) \in k[[X,Y]]$. Let us denote by $\square$ its local ring and assume that $x = X + (f)$ is a transversal parameter of $\square$.

Let

$$z_{j-1} = \sum_i a_{ji} z_j^i + z_j^{h_j} z_{j+1}, \qquad 0 \leqslant j \leqslant r,$$

be the Hamburger-Noether expansion for f, which will be assumed written in the reduced form.

Consider the values $\overline{\beta}_\nu = \overline{\beta}_\nu(C)$, $1 \leqslant \nu \leqslant g = g(\square)$, defined in the preceding section and put $\overline{\beta}_0 = n$, (these values depend only on the (a)-equisingularity class of C, (see 4.2.8.)). We shall often use the following notations:

$$N_\nu = \frac{e_{\nu-1}}{e_\nu}, \qquad 1 \leqslant \nu \leqslant g, \quad e_\nu = n_{s_\nu}.$$

$$N_0 = 1.$$

For each ν, $1 \leqslant \nu \leqslant g$, let $\Gamma_{\nu-1}$ be a curve of genus $\nu-1$ having the maximal contact with C. If $\Gamma_{\nu-1}$ is defined by the series $f_{\nu-1}(X,Y)$, we have:

$$\overline{\beta}_\nu = (C, \Gamma_{\nu-1}) = \underline{v}(f_{\nu-1}(x,y)).$$

It follows that $\overline{\beta}_\nu \in S(\square)$, $1 \leqslant \nu \leqslant g$.

Moreover, if $y = 0$ is tangent to C, then it also is to $\Gamma_{\nu-1}$, so $X + (f_{\nu-1})$ is a transversal parameter for $\Gamma_{\nu-1}$. Hence $f_{\nu-1}(X,Y)$ may be considered to be a polynomial of degree $N_1 . N_2 \ldots N_{\nu-1} = \frac{n}{e_{\nu-1}}$ in Y with coefficients in $k((X))$ (4.2.11.3.).

The following results aim to get more appropriate expressions for the $\overline{\beta}_\nu$'s. The algorithm 3.4.14. will be useful in order to compute these values.

Lemma 4.3.3.- Let m/n a rational fraction, $m > n > 0$, and assume that $c_0, c_1, \ldots, c_s$ are the partial quotients of its representation as a continued fraction. Let $r_1, \ldots, r_s$ $(r_s = (m,n))$ be the corresponding residues. If $r_0 = n$, then

$$A_s = \sum_{j=0}^{s} c_j . r_j^2 = m.n \quad .$$

Proof: By induction on s. For $s=1$, the result is evident. If it is true for $s-1$, by considering the rational fraction n/r_1, we have:

$$A_{s-1} = \sum_{j=1}^{s} c_j . r_j^2 = n.r_1 \quad .$$

Thus, $A_s = c_0 . n^2 + A_{s-1} = c_0 . n^2 + n.r_1 = n.m$, which completes the proof.

Proposition 4.3.4.- If $(e_\nu, m_\nu)_{0 \leqslant \nu \leqslant g-1}$ are as in chapter III (algorithm 3.4.14.), then

$$\overline{\beta}_\nu = \frac{1}{e_{\nu-1}} (e_0 . m_0 + \ldots + e_{\nu-1} . m_{\nu-1}) , \qquad 1 \leqslant \nu \leqslant g.$$

Proof: According to 4.2.8., we have,

$$\overline{\beta}_\nu = \frac{1}{e_{\nu-1}} \sum_{j=0}^{s_{\nu-1}} h_j . n_j^2 + n_{s_{\nu-1}+1} \quad , \quad 1 \leqslant \nu \leqslant g .$$

Hence, by using the preceding lemma, we obtain:

$$(1) \qquad e_{\nu}\overline{\beta}_{\nu+1} - e_{\nu-1}\overline{\beta}_{\nu} = \Big(\sum_{j=s_{\nu-1}+1}^{s_{\nu}-1} h_j n_j^2 + k_{\nu} n_{s_{\nu}}^2 \Big) +$$

$$+ (h_{s_{\nu}} - k_{\nu}) n_{s_{\nu}}^2 + n_{s_{\nu}} . n_{s_{\nu}+1} - n_{s_{\nu-1}} . n_{s_{\nu-1}+1} = e_{\nu} . m_{\nu} \ .$$

Now, from (1), by an inductive method, the formulae in the proposition may be obtained.

<u>Proposition 4.3.5.</u>- Let $\square$ be a curve with genus g and $(\beta_{\nu})_{0 \leq \nu \leq g}$ (resp. $(\overline{\beta}_{\nu})_{0 \leq \nu \leq g}$) the characteristic exponents (resp. the values of the maximal contact) of $\square$. Then,

$$\overline{\beta}_0 = \beta_0 ,$$

$$\overline{\beta}_{\nu} = N_{\nu-1} \cdot \overline{\beta}_{\nu-1} - \beta_{\nu-1} + \beta_{\nu} , \qquad 1 \leq \nu \leq g.$$

<u>Proof:</u> For $\nu=0$ and $\nu=1$, it is evident.

If $\nu \geq 2$, we have (see above proposition)

$$e_{\nu-1} . \overline{\beta}_{\nu} = e_{\nu-2} . \overline{\beta}_{\nu-1} + e_{\nu-1} . m_{\nu-1} .$$

Hence, by 3.4.15.,

$$\overline{\beta}_{\nu} = N_{\nu-1} \cdot \overline{\beta}_{\nu-1} - \beta_{\nu-1} + \beta_{\nu} \ .$$

<u>Corollary 4.3.6.</u>- Let $(\overline{\beta}_{\nu})_{0 \leq \nu \leq g}$ the system of values of the maximal contact for the curve $\square$. Then:

(i) $\overline{\beta}_{\nu} \in S(\square)$, $0 \leq \nu \leq g$.

(ii) $\overline{\beta}_0 < \overline{\beta}_1 < \ldots < \overline{\beta}_g$.

(iii) $(\overline{\beta}_0, \ldots, \overline{\beta}_{\nu}) = (\beta_0, \ldots, \beta_{\nu}) = e_{\nu}$, $0 \leq \nu \leq g$. In particular, $\overline{\beta}_0 > (\overline{\beta}_0, \overline{\beta}_1) > \ldots > (\overline{\beta}_0, \ldots, \overline{\beta}_g) = 1$.

Proof: (i) and (ii) are evident. We shall prove (iii) by induction on ν. For $\nu=0$, it is evident. If it holds for $\nu-1$, then using the above proposition, we have

$$(\overline{\beta}_0,\ldots,\overline{\beta}_\nu) = (\beta_0,\ldots,\beta_{\nu-1},\overline{\beta}_\nu) = (\beta_0,\ldots,\beta_\nu) = e_\nu .$$

Corollary 4.3.7.- Two plane curves are (a)-equisingular if and only if they have the same genus and the same values of the maximal contact, (i.e., the values of the maximal contact are a complete system of invariants for the equiresolution).

Proof: The formulae in 4.3.5. can be viewed as inversion formulae to compute $\overline{\beta}_0,\ldots,\overline{\beta}_g$ in terms of $\beta_0,\ldots,\beta_g$ and conversely (Note that $N_{\nu-1}$ may be obtained from $\overline{\beta}_0,\ldots,\overline{\beta}_{\nu-1}$ (or from $\beta_0,\ldots,\beta_{\nu-1}$)). The proof is therefore evident.

Theorem 4.3.8.- Let $\square$ be a curve with $(\overline{\beta}_\nu)_{0\leq\nu\leq g}$ as values of the maximal contact. Let $\{x,y\}$ a basis of the maximal ideal of $\square$, x a transversal parameter, and $\underline{v}$ the natural valuation associated to $\square$.

If $z\in\square$ and $z=g(y)$, where g is a polynomial in y of degre d $<N_1\ldots N_\gamma$, then

$$\underline{v}(z)\in\sum_{\nu=0}^{\gamma}\overline{\beta}_\nu Z_+ .$$

In particualar,

$$S(\square) = \sum_{\nu=0}^{g}\overline{\beta}_\nu Z_+ .$$

Proof: Denote by C the embedded algebroid curve defined by the basis $\{x,y\}$. We may suppose, without loss of generality, that the regular curve $y=0$ has the maximal contact with C; otherwise by a change such as $\overline{y} = y - a_{01}x - \ldots - a_{0h}x^h$, the curve $\overline{y}=0$ has the maximal contact and z can be expressed as a polynomial in $\overline{y}$

of the same degree as g.

We shall prove the theorem by induction on γ.

If $\gamma = 1$, we have

$$z = \sum_{i=0}^{d} A_i(x)\, y^i \quad , \quad d < N_1 .$$

Since $\underline{v}(A_i(x)\, y^i) \in \overline{\beta}_0 \mathbf{Z}_+ + \overline{\beta}_1 \mathbf{Z}_+$ for each i, $0 \leq i \leq d$, it is sufficient to see that these values are pairwise different.

Indeed, if $\underline{v}(A_i(x)\, y^i) = \underline{v}(A_j(x)\, y^j)$ for indices i, j $0 \leq i, j \leq d$; we have $(i-j)\overline{\beta}_1 = 0 (\mathrm{mod}.\, n)$. Hence $(i-j) = 0$ (mod. N_1), and since $|i-j| < N_1$, we must have $i = j$.

Now, assume that the result is true for $\gamma - 1$, and let $z \in \square$, $z = g(y)$, where g is a polynomial in y of degree $d < N_1 \ldots N_\gamma$. Choose a curve which has the maximal contact of genus $\gamma - 1$ with C. By 4.3.2. it can be defined by a polynomial $f_{\gamma-1}(X, Y) \in k[[X]][Y]$ of degree $N_1 \ldots N_{\gamma-1}$.

Since $f_{\gamma-1}$ is monic, we may obtain

$$g(Y) = B_o(Y) + C_1(Y)\, f_{\gamma-1}(X, Y),$$

with $B_o(Y)$, $C_1(Y) \in k[[X]][Y]$ and $\mathrm{degree}(B_o(Y)) < N_1 \ldots N_{\gamma-1}$.

By repeating the above division with $C_1(Y)$ instead of $g(Y)$ and continuing in the same way, we have

$$(1) \qquad g(Y) = \sum_{i=0}^{s} B_i(Y)\, (f_{\gamma-1}(X, Y))^i ,$$

with $B_i(Y) \in k[[X]][Y]$ and $\mathrm{degree}(B_i(Y)) < N_1 \ldots N_{\gamma-1}$. Furthermore, we have $s < N_\gamma$ since $d < N_1 \ldots N_\gamma$.

By the induction hypothesis,

$$v(B_i(y)) \in \sum_{\nu=0}^{\gamma-1} \overline{\beta}_\nu \mathbf{Z}_+ .$$

As above we must prove that these values are pairwise different.

If $\underline{v}(B_i(x)) = \underline{v}(B_j(x))$ for indices i,j, $0 \leqslant i,j \leqslant s$, recalling that $\underline{v}(f_{\gamma-1}(x,y)) = \overline{\beta}_\gamma$, we have $(i-j)\,\overline{\beta}_\gamma = 0$ (mod. $e_{\gamma-1}$) and so $(i-j) \equiv 0$ (mod. N_γ). Since $|i-j| < N_\gamma$ it follows that $i=j$, and this completes the proof.

<u>Remark 4.3.9.-</u> Let C be a plane algebroid curve with local ring □. Let $\{x,y\}$ the basis of its maximal ideal, with x a transversal parameter.

If $f_\gamma(X,Y)$ is a polynomial in Y which defines a curve having the maximal contact with C of genus γ $(0 \leqslant \gamma \leqslant g-1)$ then every $z \in$ □ has an expression of type

$$(2) \qquad z = \sum_{\substack{0 \leqslant i_\gamma < N_\gamma \\ 1 \leqslant \gamma \leqslant g}} A_{i_1,\dots,i_g}(x)\, f_0(x,y)^{i_1} \cdots f_{g-1}(x,y)^{i_g}$$

In fact, it suffices to use induction on γ, applying an appropriate induction hypothesis to each $B_i(Y)$ in (1) of the above theorem.

Furthermore, the $\underline{v}$-value of the general term in the sum (2) is $i_0\,\overline{\beta}_0 + i_1\,\overline{\beta}_1 + \dots + i_g\,\overline{\beta}_g$, where $i_0 \geqslant 0$ and $i_\gamma < N_\gamma$ for $1 \leqslant \gamma \leqslant g$. Since each two of this values are different the expression (2) for z is unique. It follows that the set

$$\{f_0^{i_1} \cdots f_{g-1}^{i_g}\}_{\substack{i_\gamma < N_\gamma \\ 1 \leqslant \gamma \leqslant g}}$$

is a basis of the free $k((x))$-module □.

<u>Corollary 4.3.10.-</u> Let □ be an irreducible plane curve with semigroup of values $S($□$)$, and having $(\overline{\beta}_\nu)_{0 \leqslant \nu \leqslant g}$ as system of

values of the maximal contact. Then we have:

(a) $\overline{\beta}_0$ is the minimum of $S(\square) - \{0\}$.

(b) For each γ, $1 \leq \gamma \leq g$, $\overline{\beta}_\gamma$ is the least integer in $S(\square)$ which does not belong to $\sum_{\nu=0}^{\gamma-1} \overline{\beta}_\nu \mathbb{Z}_+$, (i.e., $\{\overline{\beta}_0, \ldots, \overline{\beta}_g\}$ is the minimal set of generators of the semigroup $S(\square)$ (see 5.1.1.)).

(c) For each γ, $1 \leq \gamma \leq g$, $\overline{\beta}_\gamma$ is the least integer β in $S(\square)$ such that $(\overline{\beta}_0, \ldots, \overline{\beta}_{\gamma-1}, \beta) < (\overline{\beta}_0, \ldots, \overline{\beta}_{\gamma-1})$.

Proof: (a) It is evident.

(b) If $\beta \in S(\square)$ and $\beta \notin \sum_{\nu=0}^{\gamma-1} \overline{\beta}_\nu \mathbb{Z}_+$, then

$$\beta = a_0 \overline{\beta}_0 + a_1 \overline{\beta}_1 + \ldots + a_g \overline{\beta}_g , \qquad a_\nu \geq 0,$$

where $a_\nu \neq 0$ for some $\nu > \gamma - 1$. So $\beta \geq \overline{\beta}_\gamma$ and since $\overline{\beta}_\gamma \notin \sum_{\nu=0}^{\gamma-1} \overline{\beta}_\nu \mathbb{Z}_+$ (4.3.6.(c)) the result follows trivially.

(c) It derives from (b), since

$$(\overline{\beta}_0, \ldots, \overline{\beta}_\gamma) < (\overline{\beta}_0, \ldots, \overline{\beta}_{\gamma-1}).$$

Theorem 4.3.11.- Two plane curves defined over the same field are (a)-equisingular if and only if their respective semigroups of values agree.

Proof: If we recall that the values of the maximal contact are a complete system of invariants for the (a)-equisingularity, the necessity is a trivial consequence of 4.3.8. , and the suffucience follows from the above corollary.

Proposition 4.3.12.- Two given plane curves, defined over fields which are different in general, have the same semigroup of values if and only if they have a complex model in common. In particular, the semigroup of values of a curve and that of any of its models agree.

Proof: According to 4.3.5. the values of the maximal contact depend only on the desingularization process (1.5.10.). Then, the proposition derives from the above theorem applied to the complex field.

4. THE DEGREE OF THE CONDUCTOR OF $\overline{\square}$ IN $\square$.

In Chapters III and IV several systems of invariants for (a)-equisingularity have been introduced: multiplicity sequence of the desingularization process, characteristic exponents, Newton coefficients, values of the maximal contact, and semigroup of values.

All these systems remain constant when a given curve is replaced by a complex model for it, (see 3.2.3.,3.2.8.,4.1.8., 4.3.5., and 4.3.12.). The substitution of a curve for a complex model is sometimes a successful method to extend certain well known properties of the complex curves to the case of curves over arbitrary fields. As a example, we study in this section the conductor of the integral closure of a plane curve.

Consider an irreducible plane algebroid curve $\square$ over an algebraically closed field k. Let $\overline{\square} = k((t))$ the integral closure of $\square$ in its quotient field.

The conductor of $\overline{\square}$ in $\square$ is defined (see Zariski-Samuel, [27]) to be the set

$$\mathfrak{C} = \{ z \in \overline{\square} \;/\; z\overline{\square} \subset \square \}.$$

If $\{z_1,\dots,z_q\}$ is a set of generators of $\overline{\square}$ as $\square$-module (which is noetherian), and if we put $z_i = (x_i/y_i)$, $x_i, y_i \in \square$, $y_i \neq 0$, $1 \leqslant i \leqslant q$, then $y = y_1 \dots y_q \in \mathfrak{C}$ and $y \neq 0$. It follows that $\mathfrak{C} \neq 0$.

On the other hand $\mathfrak{C}$ is an ideal of $\square$ and of $\overline{\square}$ simultaneously. Thus, in particular, since $\overline{\square}$ is a discrete valuation ring with t as a uniformizing, we have $\mathfrak{C} = t^c\,\overline{\square}$ for some $c > 0$.

The integer c is called <u>the degree of the conductor of $\overline{\square}$ in $\square$</u>. Sometimes, c is said to be the <u>degree of the conductor of the integral closure of the curve.</u>

The present section, as we have just pointed out, is devoted to obtain the main properties of the integer c, which will generalize the corresponding ones in the characteristic zero case.

<u>Proposition 4.4.1.-</u> c is the only integer > 0 for which

$$(1) \qquad t^c\,\overline{\square} \subset \square \quad \text{and} \quad t^{c-1}\,\overline{\square} \not\subset \square .$$

<u>Proof:</u> If $c' > 0$ is an integer such that $t^{c'}\,\overline{\square} \subset \square$ and $t^{c'-1}\,\overline{\square} \not\subset \square$, then $t^{c'} \in \mathfrak{C}$ and $t^{c'-1} \notin \mathfrak{C}$. The first condition implies $c' \geqslant c$ and the second one implies $c'-1 < c$, so $c = c'$. Finally we note that c verifies trivially both properties in the proposition.

<u>Proposition 4.4.2.-</u> Let $S(\square)$ be the semigroup of values of a plane curve $\square$. Then c is the least integer for which

$$(2) \qquad i \in \mathbf{Z}_+ \;,\quad i \geqslant c \implies i \in S(\square).$$

<u>Proof:</u> From the above proposition c verifies (2).

Now, let c' be the least integer for which (2) is true. It is evident that $c' \leqslant c$. We shall prove $c' = c$. Indeed, if $c' < c$, then $\underline{v}(z) = c-1$ for some $z \in \square$. Then for any $w \in \overline{\square}$, $w = w_o + w'$, with $w_o \in k$ and $\underline{v}(w') \geqslant 1$, it would satisfy

$$z\,w = z\,w_o + z\,w' \in \square ,$$

so $z \in \mathfrak{C} = t^c\, \overline{\square}$, whence $\underline{v}(z) \geq c$, and we would get a contradiction.

<u>Corollary 4.4.3.</u>- The degree of the conductor c of $\overline{\square}$ in $\square$ depends only on the (a)-equisingularity class of $\square$.

<u>Corollary 4.4.4.</u>- Let $\square$ be an irreducible plane algebroid curve over k and $\square_{\mathbb{C}}$ a complex model for it. Then the degrees of the respective conductors for $\square$ and $\square_{\mathbb{C}}$ agree.

The proof of this corollary is trivial from 4.3.11. and 4.3.12.

<u>Proposition 4.4.5.</u>- Let c the degree of the conductor of the integral closure of a plane curve $\square$. The following statements hold:

(i) If $\square = \square_o \subset \ldots \subset \square_M = \overline{\square}$ is the sequence of desingularization 1.5.10. and if $\bar{e}_i = e(\square_i)$, $0 \leqslant i \leqslant M$, then

$$c = \sum_{i=0}^{M} \bar{e}_i\,(\bar{e}_i - 1) \; .$$

(ii) If $(\beta_\nu)_{0 \leqslant \nu \leqslant g}$ are the characteristic exponents of $\square$, then

$$c = (e_{g-1} - 1) + \beta_{g-1}(e_{g-2} - e_{g-1}) + \ldots + \beta_1(n - e_1) - (n-1),$$

where $e_\nu = (\beta_0, \beta_1, \ldots, \beta_\nu)$.

(iii) With notations as in section 2,

$$c = N_g\,\overline{\beta}_g - \beta_g - (n-1).$$

(iv) If $(e_\nu, m_\nu)_{0 \leqslant \nu \leqslant g-1}$ are the values obtained from the algorithm 3.4.14., then

$$c = \sum_{\nu=0}^{g-1} m_\nu (e_\nu - 1) - (n-1)$$

Proof: To prove (i), (ii), and (iii) it suffices to replace $\square$ by a complex model $\square_{\mathbb{C}}$. In fact, since for both curves the semigroup as well as the characteristic exponents and the values of the maximal contact agree, the formulae are the same as those for complex curves (see Zariski, [26]).

(iv) proceeds trivially form (iii), the theorem 3.4.15. and the proposition 4.3.4.

Proposition 4.4.6.- If c is the degree of the conductor of $\overline{\square}$ in $\square$, then for each i, $0 \leq i \leq c-1$, one and only one of the two integers i and $c-1-i$ belongs to $S(\square)$. (i.e., the semigroup $S(\square)$ is symmetric).

Proof: It is sufficient to consider the complex case (see Zariski, [26], page 28).

Proposition 4.4.7.- The ring $\square$ of an irreducible plane algebroid curve is a Gorenstein ring, i.e.,

$$l(\overline{\square}/\square) = l(\square/\mathfrak{c})$$

($\overline{\square}/\square$ and $\square/\mathfrak{c}$ considered as $\square$-modules).

Proof: Since any $\square$-submodule of $\overline{\square}/\square$ is also a k-vector space, we have $l(\overline{\square}/\square) \leq \dim_k(\overline{\square}/\square)$. In the same way $l(\square/\mathfrak{c}) \leq \dim_k(\square/\mathfrak{c})$, and $l(\overline{\square}/\mathfrak{c}) \leq \dim_k(\overline{\square}/\mathfrak{c})$.

Now, as $\{1+\mathfrak{c}, t+\mathfrak{c}, \ldots, t^{c-1}+\mathfrak{c}\}$ is a basis of the k-vector space $\overline{\square}/\mathfrak{c}$ we have $\dim_k(\overline{\square}/\mathfrak{c}) = c$. Actually $l(\overline{\square}/\mathfrak{c}) = c$, since

$$(0) \subset t^{c-1}\, \overline{\square}/\mathfrak{p} \subset \ldots \subset t\, \overline{\square}/\mathfrak{p} \subset \overline{\square}/\mathfrak{p}$$

is a strict chain of $\square$-modules.

On the other hand, $\dim_k(\square/\mathfrak{p}) = c/2$ since a basis of $\square/\mathfrak{p}$ as k-vector space may be obtained choosing any set formed by $c/2$ elements in $\square$ having respectively as $\underline{v}$-values the integers in $S(\square)$ which are less than c.

Finally, taking into account that

$$\dim_k(\overline{\square}/\square) = \dim_k(\overline{\square}/\mathfrak{p}) - \dim(\square/\mathfrak{p}) = c/2 ,$$

$$l(\overline{\square}/\square) + l(\square/\mathfrak{p}) = l(\overline{\square}/\mathfrak{p}) = c ,$$

$$l(\overline{\square}/\square) \leqslant \dim_k(\overline{\square}/\square) = c/2 ,$$

$$l(\square/\mathfrak{p}) \leqslant \dim_k(\square/\mathfrak{p}) = c/2 ,$$

the proof is trivial.

<u>Theorem 4.4.8.</u>- Let C be the embedded plane curve defined by an irreducible series

$$f(X,Y) = \sum_{i=n}^{\infty} f_i(X,Y) \in k((X,Y)) ,$$

with $n > 0$ and f_i a homogeneous polynomial of degree i. If for each $m \geqslant n$ we put

$$h_m(X,Y) = \sum_{i=n}^{m} f_i(X,Y) ,$$

and we design by C_m the curve defined by h_m, then there exists an integer $m_o > 0$, such that C and C_m are formally isomorphic for $m \geqslant m_o$.

<u>Proof:</u> By 3.4.19., for m large enough, h_m is irreducible. Let $\square$ (resp. $\square_m$) denote the local ring of C (resp. C_m). Assume that

$$z_{j-1} = \sum_i a_{ji} z_j^i + z_j^{h_j} z_{j+1}, \quad 0 \leq j \leq r,$$

is the Hamburger-Noether expansion for $\square$ in the basis $\{x, y\}$, where $x = X+(f)$ and $y = Y+(f)$.

According to theorem 3.4.18., there exists m_o such that if $m \geq m_o$, the Hamburger-Noether expansion for $\square_m$ in the basis $\{\bar{x}, \bar{y}\}$, $\bar{x} = X+(h_m)$, $\bar{y} = Y+(h_m)$ is given by

$$\bar{z}_{j-1} = \sum_i a_{ji} \bar{z}_j^i + \bar{z}_j^{h_j} \bar{z}_{j+1}, \quad 0 \leq j \leq r-1,$$

$$\bar{z}_{r-1} = a_{r2} \bar{z}_r^2 + \ldots + a_{rc} \bar{z}_r^c + g(\bar{z}_r),$$

where g is a series whose order is greater than the degree of the conductor c.

Setting $z_r = \bar{z}_r$, we may suppose that $\square \subset k((z_r))$ and $\square_m \subset k((z_r))$. If $\underline{v}$ is the natural valuation of $k((z_r))$, we have

$$\underline{v}(\bar{x}-x) \geq c$$

$$\underline{v}(\bar{y}-y) \geq c,$$

and so,

$$\bar{x} = x + g(x,y)$$

$$\bar{y} = y + g'(x,y),$$

where g and g' are series of order ≥ 2, (see 4.4.1.). Hence $\square = \square_m \subset k((z_r))$, which completes the proof.

CHAPTER V

TWISTED ALGEBROID CURVES.

This chapter treats, essentially, the problem of classifying singularities of irreducible twisted algebroid curves over an algebraically closed field. Three definitions of equisingularity, which coincide for plane curves with the (a)-equisingularity, are given.

The first one classifies singularities by means of equiresolution of the generic plane projections, the second one by equiresolution using space quadratic transformations, and the third one by the semigroup of values. However, we shall prove that in general they are different. On the other hand, none of them may be considered as a better definition than the other ones.

1. PRELIMINARY CONCEPTS AND NOTATIONS.

As in previous chapters $\mathbb{Z}_+$ is meant to denote the semigroup of nonnegative integers. Let us consider a subsemigroup $S \subset \mathbb{Z}_+$.

Definition 5.1.1.- A minimal set of generators of S is a set $\{\overline{\beta}_0, \dots, \overline{\beta}_g\} \subset \mathbb{Z}_+$ such that:

(i) $\overline{\beta}_0 > 0$.

(ii) $\overline{\beta}_{\nu+1} \notin \sum_{i=0}^{\nu} \overline{\beta}_i \mathbb{Z}_+$, $0 \leq \nu \leq g-1$, and $S = \sum_{i=0}^{g} \overline{\beta}_i \mathbb{Z}_+$.

(iii) $(\overline{\beta}_0, \overline{\beta}_1, \ldots, \overline{\beta}_g) = 1$.

According to (ii) in this definition if a minimal set of generators exists, it is unique. Thus $(\overline{\beta}_\nu)_{0\leq \nu\leq g}$ may be <u>called the minimal set of generators of S</u>. If S is the semigroup of values of a plane curve, the minimal set of generators is nothing but the system of values of the maximal contact (see 4.3.10. (b)).

<u>Proposition 5.1.2.</u>- The minimal set of generators of S exists if and only if $\mathbb{Z}_+ - S$ is finite.

<u>Proof:</u> Let $\mathbb{Z}_+ - S$ finite. Then there exists an integer $c > 0$ such that:

$$j \in \mathbb{Z}_+, \quad j \geq c \implies j \in S.$$

Set $\overline{\beta}_0 = \min(S-\{0\})$ and define inductively $\overline{\beta}_1, \ldots, \overline{\beta}_g$ by

$$\overline{\beta}_{\nu+1} = \min\left(S - \sum_{i=0}^{\nu} \overline{\beta}_i \mathbb{Z}_+\right),$$

whenever $\left(S - \sum_{i=0}^{\nu} \overline{\beta}_i \mathbb{Z}_+\right) \cap \{1,2,\ldots,c+\overline{\beta}_0\} \neq \emptyset$.

We claim that $S = \sum_{i=0}^{g} \overline{\beta}_i \mathbb{Z}_+$. In fact, let $j \in S$. If $j < c+\overline{\beta}_0$ it is evident that $j \in \sum_{i=0}^{g} \overline{\beta}_i \mathbb{Z}_+$. If $j \geq c+\overline{\beta}_0$, there exists $a'_0 \in \mathbb{Z}_+$ such that $c \leq j - a'_0\overline{\beta}_0 < c+\overline{\beta}_0$. Since $j - a'_0\overline{\beta}_0 \in S$, we have:

$$j - a'_0\overline{\beta}_0 = a_0\overline{\beta}_0 + a_1\overline{\beta}_1 + \ldots + a_g\overline{\beta}_g, \quad a_i \in \mathbb{Z}_+, \ 0\leq i\leq g.$$

Hence $j \in \sum_{i=0}^{g} \overline{\beta}_i \mathbb{Z}_+$ as desired. Moreover, it is evident that $(\overline{\beta}_\nu)_{0\leq\nu\leq g}$ is a minimal set of generators of S.

Conversely, assume that $(\overline{\beta}_\nu)_{0\leqslant\nu\leqslant g}$ is a minimal set of generators for S. Since $(\overline{\beta}_0,\ldots,\overline{\beta}_g)=1$, there exist integers $a_0,\ldots,a_g$ such that

$$a_0\,\overline{\beta}_0 + a_1\,\overline{\beta}_1 + \ldots + a_g\,\overline{\beta}_g = 1 .$$

Collecting the terms for which $a_j \geqslant 0$ (resp. $a_j < 0$) we obtain two elements $a, b \in S$ such that $b-a=1$. In other words, we have

$$a\,\mathbb{Z}_+ + (a+1)\,\mathbb{Z}_+ \subset S,$$

which implies that $\mathbb{Z}_+ - S$ is finite. This completes the proof.

Now, let $\square$ be an irreducible algebroid curve over the algebraically closed field k. Denote by $\overline{\square} = k((t))$ the integral closure of $\square$ in its quotient field and by $\underline{v}$ the natural valuation associated to $\square$. As in 4.4. the conductor $\mathfrak{C}$ of $\overline{\square}$ in $\square$ has a nonzero element, and thus $\mathfrak{C}=(t^c)\,\overline{\square}$ for an integer $c>0$, which will be called the <u>degree of the conductor of $\overline{\square}$ in $\square$</u> or <u>the degree of the conductor of the integral closure of the curve</u>.

<u>Definition 5.1.3.</u>- The <u>semigroup of values of $\square$</u> is defined to be the subsemigroup $S(\square) = \underline{v}\,(\square - \{0\})$ of $\mathbb{Z}_+$.

<u>Lemma 5.1.4.</u>- The degree of the conductor of $\overline{\square}$ in $\square$ is the least integer in $S(\square)$ such that

$$j \in \mathbb{Z}_+ ,\quad j \geqslant c \quad \Longrightarrow \quad j \in S(\square).$$

In particular, $\mathbb{Z}_+ - S(\square)$ is finite.

<u>Proof:</u> It works as in 4.4.2.

<u>Proposition 5.1.5.</u>- A subsemigroup S of $\mathbb{Z}_+$ is the semigroup

of values of a curve if and only if $S - \mathbf{Z}_+$ is finite.

Proof: Assume $S = S(\square)$ for a curve $\square$. Then $\mathbf{Z}_+ - S$ is finite according to the preceding lemma. Conversely, if $\mathbf{Z}_+ - S$ is finite it suffices to take

$$\square = k((t^{\overline{\beta}_0}, \ldots, t^{\overline{\beta}_g})),$$

where $(\overline{\beta}_\nu)_{0 \leq \nu \leq g}$ is a minimal set of generators of S.

Now, we shall denote by $\underline{m}$ the maximal ideal of $\square$, and by $f_1 < f_2 < \ldots < f_N$ the elements in $\underline{v}(\underline{m}) - \underline{v}(\underline{m}^2)$ (Note that $\underline{v}(\underline{m}) - \underline{v}(\underline{m}^2)$ is finite since if $i \geqslant c + \overline{\beta}_0$ and $\underline{v}(z) = i$ we have $z \in \underline{m}^2$).

Lemma 5.1.6.- Let $x_i \in \underline{m}$ such that $\underline{v}(x_i) = f_i$, $1 \leqslant i \leqslant N$. Then $\{x_i\}_{1 \leqslant i \leqslant N}$ is a minimal basis of $\underline{m}$. In particular, $N = \mathrm{Emb}(\square)$.

Proof: We must prove that $\{x_i + \underline{m}^2\}_{1 \leqslant i \leqslant N}$ is a basis of the k-vector space $\underline{m}/\underline{m}^2$.

First, if $\sum_{i=1}^{N} a_i x_i \in \underline{m}^2$, with $a_i \in k$, we have

$$\underline{v}\Big(\sum_{i=1}^{N} a_i x_i\Big) \in \underline{v}(\underline{m}^2) ,$$

so $a_i = 0$ for all i, $1 \leqslant i \leqslant N$. Thus $\{x_i + \underline{m}^2\}_{1 \leqslant i \leqslant N}$ are lineary independent.

They are also generators. In fact, for any $z \in \underline{m} - \underline{m}^2$, either $\underline{v}(z) \in \underline{v}(\underline{m}^2)$ or $\underline{v}(z) \notin \underline{v}(\underline{m}^2)$. If $\underline{v}(z) \in \underline{v}(\underline{m}^2)$, there exists $z_1 \in \underline{m}^2$ such that, if $z^* = z - z_1$, we have $\underline{v}(z^*) \in \underline{v}(\underline{m}^2)$ and $\underline{v}(z^*) > \underline{v}(z)$. If $\underline{v}(z) \notin \underline{v}(\underline{m}^2)$, then $\underline{v}(z) = \underline{v}(x_i)$ for some i, $1 \leqslant i \leqslant N$; so $\underline{v}(z - a_i x_i) > \underline{v}(z)$ for $a_i \in k$. It follows that

$$z = \sum_{i=1}^{N} a_i z_i + z' , \qquad \text{with } a_i \in k \text{ and } z' \in \underline{m}^2 ,$$

whence $z + \underline{m}^2 = \sum_{i=1}^{N} a_i (x_i + \underline{m}^2)$ as desired.

Proposition 5.1.7.- We keep the notations as above. If $(\overline{\beta}_\nu)_{0 \leq \nu \leq g}$ is the minimal set of generators of $S(\square)$, we have

$$f(\square) = \{f_1, f_2, \ldots, f_N\} \subset \{\overline{\beta}_0, \overline{\beta}_1, \ldots, \overline{\beta}_g\}$$

In particular $\mathrm{Emb}(\square) \leq g+1$.

Proof: Take $f_j \in f(\square)$, and let i be the greatest integer for which $\overline{\beta}_i \leq f_j$. Assume $\overline{\beta}_i < f_j$. Since $f_j \in S(\square)$, we have

$$f_j = \sum_{\nu=0}^{i} a_\nu \overline{\beta}_\nu \;, \quad a_\nu \in Z_+ \;, \quad \sum_{\nu=0}^{i} a_\nu \geq 2 \;.$$

Take $y_\nu \in \underline{m}$ such that $\underline{v}(y_\nu) = \overline{\beta}_\nu$. Then $f_j = \underline{v}(y_o^{a_o} \ldots y_i^{a_i}) \in \underline{v}(\underline{m}^2)$ which is a contradiction. Thus $f_j = \overline{\beta}_i$.

Corollary 5.1.8.- Let $y_\nu \in \underline{m}$ be an element whose order is $\overline{\beta}_\nu$, $0 \leq \nu \leq g$. Then $\overline{\beta}_\nu \in f(\square)$ if and only if there is no polynomial $p(Y_0, \ldots, Y_{\nu-1})$ without linear terms, such that

$$\underline{v}(\,p(y_0, \ldots, y_{\nu-1})) = \overline{\beta}_\nu \;.$$

Proof: Necessity is obvious. Conversely, let $\overline{\beta}_\nu \in f(\square)$. Since $\overline{\beta}_\nu \notin \underline{v}(\underline{m}^2)$ and $\{y_0, \ldots, y_g\}$ is a basis of $\underline{m}$, there exists a polynomial $q(Y_0, \ldots, Y_g)$ without linear terms, such that

$$\underline{v}(\,q(y_0, \ldots, y_g)\,) = \overline{\beta}_\nu \;.$$

If $p(Y_0, \ldots, Y_{\nu-1})$ is the polynomial obtained from $q(Y_0, \ldots, Y_g)$ by collecting all its monomials in $Y_0, \ldots, Y_{\nu-1}$, we have

$$\underline{v}(p(y_0, \ldots, y_{\nu-1})) = \underline{v}(q(y_0, \ldots, y_g)) = \overline{\beta}_\nu \;.$$

Corollary 5.1.9.- $f_1 = \overline{\beta}_0$ and $f_2 = \overline{\beta}_1$ (if $g > 1$). Moreover, $\square$ is plane if and only if $f(\square) = \{\overline{\beta}_0\}$ or $f(\square) = \{\overline{\beta}_0, \overline{\beta}_1\}$.

Remark 5.1.10.- If $S = \sum_{\nu=0}^{g} \overline{\beta}_\nu \mathbb{Z}_+$ (where $(\overline{\beta}_\nu)_{0 \leq \nu \leq g}$ the minimal set of generators of S) and $f = \{f_1, \dots, f_N\} \subset \{\overline{\beta}_0, \dots, \overline{\beta}_g\}$ with $g > 1$, $f_1 = \overline{\beta}_0$, and $f_2 = \overline{\beta}_1$, it is not true that $f = f(\square)$ and $S = S(\square)$ for some curve $\square$. For instance if S is not symmetric (see 4.4.6.), there is no plane curve such that $S(\square) = S$.

2. EQUISINGULARITY E.s.1. GENERIC PLANE PROJECTIONS.

Consider a singular irreducible algebroid curve $\square$ over k. In this section $\overline{\square}$ continues to denote the integral closure of $\square$ in its quotient field, $\underline{v}$ the natural valuation of $\square$ and $S(\square) = \sum_{\nu=0}^{g} \overline{\beta}_\nu \mathbb{Z}_+$ the semigroup of values of $\square$, where $(\overline{\beta}_\nu)_{0 \leq \nu \leq g}$ the minimal set of generators of $S(\square)$.

Let $P(\square)$ denote the set formed by the algebroid curves $\square'$ such that $k \subset \square' \subset \square$ and $\mathrm{Emb}(\square') \leq 2$. For such a curve $\square'$ we have

$$\underline{v}(\square' - \{0\}) \subset \underline{v}(\square - \{0\}) = S(\square) \subset \mathbb{Z}_+ .$$

Since infinite subsets of $\mathbb{Z}_+$ may be viewed as strictly increasing sequences in $\mathbb{Z}_+$, the set

$$S(P(\square)) = \{ \underline{v}(\square' - \{0\}) \mid \square' \in P(\square) \} \subset \mathbb{Z}_+^{\mathbb{Z}_+}$$

may be ordered by the order $\lessdot$ induced by the lexicographic order of $\mathbb{Z}_+^{\mathbb{Z}_+}$.

Lemma 5.2.1.- $(S(P(\square)), \preccurlyeq)$ is a well ordered set.

Proof: Let M be a nonempty subset of $S(P(\square))$. We may construct, by induction, a sequence of integers $m_1 < m_2 < \dots < m_i < \dots$ as follows:

First, we define $m_1 = \min\{ t \mid t = \min S' \text{ for some } S' \in M\}$ and we put $M_1 = \{S' \in M \mid m_1 = \min S'\}$.

In the inductive step, if $m_1, \dots, m_i$ and $M_i \subset M$, $M_i \neq \emptyset$, are constructed, we define

$$m_{i+1} = \min\{ t \mid t = \min(S' - \{m_1, \dots, m_i\}) \text{ for some } S' \in M_i \}$$

and we put $M_{i+1} = \{ S' \in M_i \mid m_{i+1} = \min (S' - \{m_1, \dots, m_i\})\}$.

Now, by the above construction the set $S^* = \{m_i \mid 1 \leqslant i < \infty \}$ is the infimum of M in $\mathbb{Z}_+^{\mathbb{Z}_+}$ for the lexicographic order. We must see that $S^* \in M$. To prove this, note that for any $r > 0$, if i is an index such that $m_i > r$ and if $S' \in M_i$, we have $S^* \cap \{1, \dots, r\} = S' \cap \{1, \dots, r\}$. It follows that S^* is a semigroup (if $m, m' \in S^*$, taken $r > m+m'$ and S' as above, we have $m+m' \in S' \cap \{1, \dots, r\}$ and hence $m+m' \in S^*$). Furthermore, if $S^* = \sum_{\nu=0}^{g} \overline{\beta}_\nu^* \mathbb{Z}_+$, where,

$$\overline{\beta}_0^* = \min (S^* - \{0\}),$$

$$\overline{\beta}_{\nu+1}^* = \min \left(S^* - \sum_{i=0}^{\nu} \overline{\beta}_i^* \mathbb{Z}_+ \right),$$

r may be chosen such that $r > \overline{\beta}_g^*$. Thus, since $\overline{\beta}_\nu^* \in S'$, $0 \leqslant \nu \leqslant g$, we have $S^* \subset S'$, whence, $S' \preccurlyeq S^*$, and therefore $S^* = S' \in M$.

Notation 5.2.2.- The minimum element for $\preccurlyeq$ in $S(P(\square))$ will be denoted along this section by $S^*(\square)$.

Definition 5.2.3.- A curve $\square' \in P(\square)$ will be called a generic plane projection of $\square$ when $S^*(\square) = \underline{v}(\square' - \{0\})$.

Proposition 5.2.4.- $\square$ is plane if and only if $S(\square) = S^*(\square)$.

Proof: If $\square$ is plane we have $\square \in P(\square)$ and so $S^*(\square) \leqslant S(\square)$. On the other hand since $S^*(\square) \subset S(\square)$ it is evident that $S(\square) \leqslant S^*(\square)$.

Now assume that $S^*(\square) = S(\square)$ and let $\square'$ be a generic plane projection of $\square$. We claim that $\square' = \square$. In fact, for any $z \in \square$, we may construct inductively a sequence $z_1, z_2, \ldots, z_m, \ldots$ with $z_i \in \square'$, such that

$$\lim_{m \to \infty} \underline{v}(z - (z_1 + \ldots + z_m)) = \infty .$$

Hence, since $\square'$ is a closed subset of $\square$, we have $z = \sum_{i=0}^{\infty} z_i \in \square'$, which completes the proof.

Lemma 5.2.5.- Let $\square'$ be a generic plane projection of $\square$, then:

(i) $\mathrm{Emb}(\square') = 2$.

(ii) $\beta_0(\square') = \overline{\beta}_0$ and $\beta_1(\square') = \overline{\beta}_1$ (β_0, β_1 are characteristic exponents).

Proof: To see (i) assume that $\square' = k((z))$, $z \in \square$. Then $S^*(\square) = \underline{v}(z)\,\mathbf{Z}_+$. If we consider $x, y \in \square$ such that $\underline{v}(x) = \overline{\beta}_0$ and $\underline{v}(y) = \overline{\beta}_1$, the inequality $S(\square') \leqslant \underline{v}(k((x,y)) - \{0\})$ is a contradiction. Thus $\mathrm{Emb}(\square') = 2$.

Now, let $\{x_1, x_2\}$ be a basis of the maximal ideal of $\square'$ such that $\underline{v}(x_1)$ does not divide $\underline{v}(x_2)$ and $\underline{v}(x_1) < \underline{v}(x_2)$. As above

$$S(\square') \leqslant \underline{v}(k((x,y)) - \{0\})$$

implies easily $\underline{v}(x_1) = \overline{\beta}_0$ and $\underline{v}(x_2) = \overline{\beta}_1$.

Lemma 5.2.6.- $\mathbf{Z}_+ - S^*(\square)$ is finite, i.e., any uniformizing parameter t for $\square$ is also a uniformizing parameter for any generic

plane projection.

<u>Proof</u>: Let $\square'$ be a generic plane projection of $\square$, and $\{x_1, x_2\}$ a basis of its maximal ideal. Assume that the Hamburger-Noether expansion in that basis is given by

$$(D') \qquad z_{j-1} = \sum_i a_{ji} z_j^i + z_j^{h_j} z_{j+1} \ , \quad 0 \leq j \leq r.$$

According 2.2.6. we must prove $\underline{v}(z_r) = 1$. Otherwise, if

$$z_r = b_s t^s + b_{s+1} t^{s+1} + \dots$$

with $b_j \in k$ and $s > 1$, we may consider the Hamburger-Noether expansion given by

$$\bar{z}_{j-1} = \sum_i a_{ji} \bar{z}_j^i + \bar{z}_j^{h_j} \bar{z}_{j+1} \ , \quad 0 < j < r-1,$$

$$(D'') \qquad \bar{z}_{r-1} = a_{r2} z_r^2 + \dots + a_{rc} z_r^c + z_r^c t$$

$$z_r = b_s t^s + b_{s+1} t^{s+1} + \dots$$

where c is the degree of the conductor of $\overline{\square}$ in $\square$. Since $\underline{v}(\bar{z}_0 - z_0) \geq c$ and $\underline{v}(\bar{z}_{-1} - z_{-1}) \geq c$, we have $\square'' = k((\bar{z}_0, \bar{z}_{-1})) \subset \square$ and so $\square'' \in P(\square)$. Finally from 4.3.4. and 4.3.8. it follows that

$$\underline{v}(\square'' - \{0\}) < \underline{v}(\square' - \{0\}) = S^*(\square),$$

which is a contradiction. This completes the proof.

Now, we aim to give a geometric interpretation of generic plane projections defined above. Let us consider a generic plane projection $\square'$ of $\square$. If $\underline{m}$ is the maximal ideal of $\square$ and $\{x_1, x_2\}$ a basis of the maximal ideal of $\square'$, $x_1 + \underline{m}^2$ and $x_2 + \underline{m}^2$

are lineary independent (see 5.1.6., 5.1.9. and 5.2.5.), and therefore it may be completed to a basis $\{x_i\}_{1 \leq i \leq N}$ (even a minimal basis if we want) of $\underline{m}$. Denote by C the corresponding embedded curve, and assume that

$$(1) \qquad x_i = x_i(t) \quad , \ 1 \leq i \leq N,$$

is a parametric representation in that basis.

Let Π be the 2-plane in k^N defined by $X_3 = \ldots = X_N = 0$, and consider the open dense subset $G \subset G_0^{N,N-2}$ (= grassmannian of (N-2)-planes which pass throught the origin) formed by those planes H such that $H \cap \Pi = \{0\}$. Any (N-2)-plane of G is given by

$$(2) \qquad \begin{aligned} X_1 + \lambda_3 X_3 + \ldots + \lambda_N X_N &= 0 \\ X_2 + \mu_3 X_3 + \ldots + \mu_N X_N &= 0, \end{aligned}$$

where $(\underline{\lambda}, \underline{\mu}) = (\lambda_3, \ldots, \lambda_N, \mu_3, \ldots, \mu_N) \in k^{2N-4}$. Conversely, any equations of type (2) define an element of G which will be denoted by $H(\underline{\lambda}, \underline{\mu})$.

The parallel projection to $H(\underline{\lambda}, \underline{\mu})$ of C on the plane Π is the plane algebroid curve over k given by

$$(3) \qquad C_{(\underline{\lambda}, \underline{\mu})}: \quad \begin{aligned} X &= x_1(t) + \lambda_3 x_3(t) + \ldots + \lambda_N x_N(t) \\ Y &= x_2(t) + \mu_3 x_3(t) + \ldots + \mu_N x_N(t) \end{aligned}$$

Equations (3) may be viewed as a parametrization for a curve over an algebraic closure F of the quotient field of $A = k[\![\underline{\lambda}, \underline{\mu}]\!]$. This curve has a Hamburger-Noether expansion (over F) given by

$$(D) \qquad Z_{j-1} = \sum_i a_{ji} Z_j^i + Z_j^{h_j} Z_{j+1} \ , \ 0 \leq j \leq r.$$

Furthermore, since $X(t), Y(t) \in A((t))$, the coefficients

a_{ji} are actually in $k(\underline{\lambda},\underline{\mu})$.

Consider the localization A_m of A relative to the maximal ideal $m = (\underline{\lambda}, \underline{\mu})$. One has the ring epimorphism (evaluation map):

$$e : A_m \longrightarrow k$$

$$\frac{h(\underline{\lambda},\underline{\mu})}{h'(\underline{\lambda},\underline{\mu})} \longmapsto \frac{h(\underline{0},\underline{0})}{h'(\underline{0},\underline{0})}$$

which has a natural extension $e : A_m((t)) \longrightarrow k((t))$.

Finally, if $q_0,\ldots,q_s \in A$, we shall denote by $A_{q_0,\ldots,q_s}$ the localization of A relative to the multiplicatively closed set $\{q_0^{i_0}\cdots q_s^{i_s} \;/\; i_0,\ldots,i_s \geq 0\}$.

Lemma 5.2.7.- Keeping notations as above, there exist $p_0,\ldots,p_g \in A$, $e(p_0\cdots p_g) \neq 0$, such that:

(i) $a_{ji} \in A_{p_0,\ldots,p_g}$, $0 \leq j \leq r$, $1 \leq i \leq h_j$.

(ii) $Z_j \in A_{p_0,\ldots,p_g}((t))$, $0 \leq j \leq r$.

(iii) The leading coefficient of Z_j is a unit in $A_{p_0,\ldots,p_g}$.

(hence $\underline{v}(Z_j) = \underline{v}(e(Z_j))$).

Proof: Assume that (D) is written in the reduced form 3.3.4. Let $p_0 \in A$ be the leading coefficient of X. Since x_1 is transversal we have $e(p_0) \neq 0$. Moreover, $a_{0i} \in A_{p_0}$ and $Z_1 \in A_{p_0}((t))$. Now, let $g_1 = p_1 / p_0^{m_1}$ be the leading coefficient of Z_1. We claim that $e(g_1) \neq 0$. In fact, if $e(g_1) = 0$ the first value of the maximal contact $\overline{\beta}_1^{(0)}$ for $k((x_1,x_2)) = \square'$ would be greater than $h\,n + n_1 = \overline{\beta}_1$. Thus setting

$$\square'' = k((X(\underline{\lambda},\underline{\mu}), Y(\underline{\lambda},\underline{\mu}))),$$

where $(\underline{\lambda}, \underline{\mu}) \in k^{2N-4}$ verifies $p_0(\underline{\lambda},\underline{\mu}).p_1(\underline{\lambda},\underline{\mu}) \neq 0$, we would have $\square'' \in P(\square)$ and

$$\underline{v}(\square'' -\{0\}) < \underline{v}(\square' -\{0\}) = S^*(\square),$$

which is a contradiction.

Now, for any j, $1 \leqslant j \leqslant s_1$, it is evident that $Z_j \in A_{p_0,p_1}((t))$ and that the leading coefficient of Z_j is a unit in A_{p_0,p_1}. It follows that $a_{s_1,i} \in A_{p_0,p_1}$, $k_1 \leqslant i \leqslant h_{s_1}$, and $Z_{s_1+1} \in A_{p_0,p_1}((t))$. If $g_2 = \dfrac{p_2}{p_0^{m_2} p_1^{m'_2}}$ denotes the leading coefficient of Z_{s_1+1}, we must have $e(p_2) \neq 0$. Otherwise the second value $\overline{\beta}_2^{(0)}$ of the maximal contact for $\square'$ would be greater than that of $\square'' = k((X(\underline{\lambda}, \underline{\mu}), Y(\underline{\lambda}, \underline{\mu})))$, choosing $(\underline{\lambda},\underline{\mu})$ such that

$$p_0(\underline{\lambda},\underline{\mu}).p_1(\underline{\lambda},\underline{\mu}).p_2(\underline{\lambda},\underline{\mu}) \neq 0.$$

This is as above a contradiction.

Thus, by using induction we may construct $p_0,\ldots,p_g \in A$ with $e(p_0\ldots p_g) \neq 0$ such that $a_{ji} \in A_{p_0,\ldots,p_g}$,

$Z_j \in A_{p_0,\ldots,p_g}((t))$ and the leading coefficient of each Z_j is a unit in $A_{p_0,\ldots,p_g}$. This completes the proof.

<u>Theorem 5.2.8.</u>- There exists an open dense $\Omega \subset k^{2N-4}$ containing $(\underline{0},\underline{0})$ such that if $(\underline{\lambda}_o, \underline{\mu}_o) \in \Omega$ the curve

$$\square_{(\underline{\lambda}_o, \underline{\mu}_o)} = k((X(\underline{\lambda}_o,\underline{\mu}_o), Y(\underline{\lambda}_o,\underline{\mu}_o)))$$

is a generic plane projection of $\square$. In particular the curves $\square_{(\underline{\lambda}_o, \underline{\mu}_o)}$ with $(\underline{\lambda}_o, \underline{\mu}_o) \in \Omega$ are (a)-equisingular.

Proof: Define

$$\Omega = \{(\underline{\lambda}_o, \underline{\mu}_o) \in k^{2N-4} / p_0(\underline{\lambda}_o, \underline{\mu}_o)\ldots p_g(\underline{\lambda}_o, \underline{\mu}_o) \neq 0\},$$

where $p_0, \ldots, p_g$ are the polynomials in the previous lemma. It is evident that $(0,0) \in \Omega$. On the other hand for $(\underline{\lambda}_o, \underline{\mu}_o) \in \Omega$, $\square_{(\underline{\lambda}_o, \underline{\mu}_o)}$ and $\square_{(\underline{0},\underline{0})} = \square'$ are (a)-equisingular (since their respective Hamburger-Noether expansions,

$$\tilde{z}_{j-1} = \sum_i a_{ji} \tilde{z}_j^{\,i} + \tilde{z}_j^{\,h_j} \tilde{z}_{j+1}, \quad 0 < j \leq r,$$

$$z_{j-1} = \sum_i e(a_{ji}) z_j^{\,i} + z_j^{\,h_j} z_{j+1}, \quad 0 \leq j < r,$$

verify $\underline{v}(\tilde{z}_j) = \underline{v}(z_j)$, $0 \leq j \leq r$). Therefore:

$$\underline{v}(\square_{(\underline{\lambda}_o, \underline{\mu}_o)} - \{0\}) = \underline{v}(\square' - \{0\}) = S^*(\square).$$

Remark 5.2.9.- The semigroup $S^*(\square)$ is actually the semigroup of values of the curve over F given by (3).

Definition 5.2.10.- Two given irreducible algebroid curves $\square_1$ and $\square_2$ over k are said to be equisingular E.s.1. iff $S^*(\square_1) = S^*(\square_2)$.

Remark 5.2.11.- According to this definition and 4.3.11. equisingularity E.s.1. means equiresolution of the generic plane projections. A complete system of invariants for equisingularity E.s.1. is the minimal set of generators of $S^*(\square)$, i.e., the system of values of the maximal contact for any generic plane projection.

Moreover, since $S^*(\square)$ is the semigroup of values of a plane curve, characteristic exponents $(\beta_\nu)_{0 \leq \nu \leq g}$ are defined by $S^*(\square)$. They may be called characteristic exponents of $\square$ and actually are also a complete system of invariants for E.s.1.

We shall consider now the case caract $k = 0$ in which Equisingularity E.s.1. has a more precise description. This description, from an algebraic view point, is made in terms of saturation of algebroid curves. Therefore, we shall introduce briefly the concept of saturation, which is extensively studied in Zariski [27].

Let $\square$ be an irreducible algebroid curve over an algebraically closed field of characteristic 0. Let $\overline{\square}$ be the integral closure of $\square$ it its quotient field, and denote by $\underline{v}$ the natural valuation of $\overline{\square}$. Take a parameter y of the curve. If $m = \underline{v}(y)$ we may choose a uniformizing t such that $y = t^m$. It is trivial to check that $k((t))/_{k((y))}$ is a Galois extension whose Galois group G is cyclic generated by the k-automorphism

$$f : k((t)) \longrightarrow k((t))$$
$$t \longmapsto wt ,$$

where w is a primitive m-the root of unity. Moreover, $\overline{\square} = k[[t]]$ is invariant by all automorphisms in G.

The <u>saturation of $\square$ with respect to y</u> is defined to be the ring $\widetilde{\square}_y$ which satisfies the following conditions:

1) $\square \subset \widetilde{\square}_y \subset \overline{\square}$.

2) If $z \in \square_y$, $z' \in \overline{\square}$ and $\underline{v}(g(z')-z') \geqslant \underline{v}(g(z)-z)$ $\forall\ g \in G$, then $z' \in \widetilde{\square}_y$.

3) $\widetilde{\square}_y$ is the smallest ring satisfying 1) and 2).

In general, a ring $\square' \subset \overline{\square}$ is said to be saturated with respect to y when 2) holds for it. Thus $\widetilde{\square}_y$ is the smallest saturated ring with respect to y which contains $\square$. On the other hand, note that $\widetilde{\square}_y$ is itself an irreducible algebroid curve over k (see 1.1.5.).

We shall need an important property of the saturation (Zariski, [27]): If x and x' are transversal parameters for $\square$, then $\widetilde{\square}_x = \widetilde{\square}_{x'}$.

Now, we shall discuss the Equisingularity E.s.1. when k is an algebraically closed field of characteristic zero.

Let $\square$ be an irreducible algebroid curve over k, and $\{x_i\}_{1 \leqslant i \leqslant N}$ any basis of $\underline{m}$, where x_1 is transversal. Assume that

$$(4) \qquad \begin{aligned} x &= t^n \\ y &= \sum_{i \geqslant n}^{\infty} a_{ji}\, t^i \quad , \quad 2 \leqslant j \leqslant N, \end{aligned}$$

is a Puiseux primitive parametric representation in that basis. Let I denote the set of integers which are effective exponents of (4), i.e., $\beta \in I \iff b_{j\beta} \neq 0$ for some j or $\beta = n$.

Define $\beta'_0 = n$ and, by using induction, set

$$\beta'_{\nu+1} = \min \{ \beta \in I \,/\, (\beta'_0, \ldots, \beta'_\nu, \beta) < (\beta'_0, \ldots, \beta'_\nu) \}.$$

Since (4) is primitive, there exist $g > 0$ such that

$$\beta'_0 > (\beta'_0, \beta'_1) > \ldots . > (\beta'_0, \ldots, \beta'_g) = 1.$$

The set $(\beta'_\nu)_{0 \leqslant \nu \leqslant g}$ does not depend on the basis $\{x_i\}_{1 \leqslant i \leqslant N}$, and it is determined by $\square$. In fact, if $\tilde{\square}$ is the <u>absolute saturation of $\square$</u> (i.e., the saturation of $\square$ with respect to any transversal parameter), then $\tilde{\square}$ is an irreducible algebroid curve over k such that, if $\tilde{S}$ is its semigroup of values, we have

$$\beta'_0 = n$$

$$\beta'_{\nu+1} = \min \{ \beta \in \tilde{S} \,/\, (\beta'_0, \beta'_1, \ldots, \beta'_\nu, \beta) < (\beta'_0, \ldots, \beta'_\nu) \}, \qquad 0 \leqslant \nu \leqslant g-1.$$

(see Zariski, [27], and J.L. Vicente, [21]).

Lemma 5.2.12.- Keeping notations as above, $(\beta'_\nu)_{0 \leq \nu \leq g}$ are the characteristic exponents of the curve defined over the algebraic closure of $k(\mu_3, \ldots, \mu_N)$ by

$$X' = x_1$$

$$Y' = x_2 + \mu_3 x_3 + \ldots + \mu_N x_N .$$

Proof: It is evident, according to the definition of $(\beta'_\nu)_{0 \leq \nu \leq g}$.

Proposition 5.2.13.- The set $(\beta_\nu')_{0 \leq \nu \leq g}$ is the set of characteristic exponents of the curve $C_{(\underline{\lambda}, \underline{\mu})}$ defined over the algebraic closure of $k(\underline{\lambda}, \underline{\mu})$ by

$$(3) \qquad \begin{aligned} X &= x_1 + \lambda_3 x_3 + \ldots + \lambda_N x_N \\ Y &= x_2 + \mu_3 x_3 + \ldots + \mu_N x_N . \end{aligned}$$

Proof: There exists an open dense $\Omega' \subset K^{N-2}$ such that if $\underline{\lambda}_o \in \Omega'$ the curves $C_{(\underline{\lambda}, \underline{\mu})}$ and $C_{(\underline{\lambda}_o, \underline{\mu})}$ have the same characteristic exponents and $X(\underline{\lambda}_o) = x_1 + \lambda_{3,o} x_3 + \ldots + \lambda_{N,o} x_N$, (where $\underline{\lambda}_o = (\lambda_{3,o}, \ldots, \lambda_{N,o})$), is transversal.

Now, the proof follows from lemma 5.2.12. applied to the basis $\{X(\underline{\lambda}_o), x_2, \ldots, x_N\}$.

Proposition 5.2.14.- The semigroup defined by the characteristic exponents $(\beta'_\nu)_{0 \leq \nu \leq g}$ (see 4.3.5.) is $S^*(\square)$.

Proof: By using the above proposition $(\beta'_\nu)_{0 \leq \nu \leq g}$ are the characteristic exponents defined in remark 5.2.11. The proof is tivial.

Corollary 5.2.15.- For two given algebroid curves over an algebraically closed field of characteristic zero, the following statements are equivalent:

(a) They are equisingular E.s.1.

(b) They have the same set $(\beta'_{\nu})_{0 \leq \nu \leq g}$.

(c) Its respective absolute saturations are isomorphic.

Remark 5.2.16.- If charact. k = p > 0 and (n,p) = 1 the saturation criterion works in the same way and a result as 5.2.14. may be given (see chapter III, 4th section).

If $(n,p) \neq 1$, there are many patological cases. For instance a lemma as 5.2.12. does not hold, since if charact. k = 2, and $\square$ is given by

$$x = t^{8}$$
$$y = t^{10}$$
$$z = t^{13},$$

the characteristic exponents of

$$x = t^{8}$$
$$y = t^{10} + \mu t^{13}$$

are 8,10,31 ; but if we use the basis $\{x', y, z\}$, where $x' = t^{8} + t^{13}$, the characteristic exponents of

$$x' = t^{8} + t^{13}$$
$$y' = t^{10} + \mu t^{13}$$

are 8,10,15 . This means that a saturation criterion does not hold.

3. EQUISINGULARITY E.s.2. SPACE QUADRATIC TRANSFORMATIONS.

Our second definition of equisingularity (E.s.2.) will be the classical definition which uses space quadratic transformations. This definition was actually given in 1.5.12. and it was called equiresolution. Namely:

Definition 5.3.1.- Two irreducible algebroid curves $\square$ and $\square^*$ over k will be called equisingular E.s.2. iff $\square$ and $\square^*$ have the same process of resolution 1.5.10.

Recall that equiresolution is characterized by using Hamburger-Noether expansions (see 2.4.9.5.): If,

$$\bar{Z}_{j-1} = \sum_i A_{ji} z_j^i + Z_j^{h_j} z_{j+1} \quad , \quad 0 \leqslant j \leqslant r,$$

$$\bar{Z}^*_{j-1} = \sum_i A^*_{ji} z_j^{*i} + Z_j^{*h_j} z^*_{j+1} \quad , \quad 0 \leqslant j \leqslant r^*,$$

are respective Hamburger-Noether expansions for $\square$ and $\square^*$, then those curves are equisingular E.s.2. iff $r=r^*$, $h_j=h_j^*$, and $n_j = \underline{v}(Z_j) = \underline{v}(Z_j^*) = n_j^*$, $0 \leqslant j \leqslant r$.

Remark 5.3.2.- The values h_j, n_j $(0 \leqslant j \leqslant r)$, in the Hamburger-Noether expansion must satisfy some requirements:

(i) $n > n_1 > \ldots > n_r = 1$, $h_j \geqslant 1$, $0 \leqslant j \leqslant r$.

(ii) If $n_j \nmid n_{j-1}$, then

$$h_j \leqslant \left[\frac{n_{j-1}}{n_j}\right] \quad \text{and} \quad n_{j+1} \leqslant n_{j-1} - h_j n_j$$

(iii) Furthermore, if

$$n_{j+\alpha} \nmid n_{j-1} - h_j n_j - \ldots - h_{j+\alpha-1} n_{j+\alpha-1}, \quad 0 \leq \alpha \leq s,$$

then

$$h_{j+s} \leq \left[\frac{n_{j-1} - h_j n_j - \ldots - h_{j+s-1} n_{j+s-1}}{n_{j+s}}\right]$$

and

$$n_{j+s+1} \leq n_{j-1} - h_j n_j - \ldots - h_{j+s} n_{j+s}.$$

Conversely, one can prove that if integers h_j, n_j $(0 \leq j \leq r)$ verify (i), (ii), and (iii), then there exists a curve over k such that these integers are actually those of any of its Hamburger-Noether expansions.

However, if an embedding dimension N is given, we are not able in general to state the existence of a curve $\square$, with $\mathrm{Emb}(\square) = N$ and with the values h_j, n_j $(0 \leq j \leq r)$ in its Hamburger-Noether expansions. Indeed, for $N=2$ we have the following evident result:

Proposition 5.3.3.- Keeping the notations as above, if $\square$ is a plane curve, then:

(i) $n > n_1 > \ldots > n_r = 1$, $h_j \geq 1$, $0 \leq j \leq r$.

(ii) If $n_j \nmid n_{j-1}$ then

$$h_j = \left[\frac{n_{j-1}}{n_j}\right] \quad \text{and} \quad n_{j+1} = n_{j-1} - h_j n_j .$$

(iii) If $n_j \mid n_{j-1}$, then

$$h_j \geq \frac{n_{j-1}}{n_j} .$$

Now, we shall give examples in order to check that between E.s.1. and E.s.2. there is no relation.

<u>Examples 5.3.4.-</u> Assume $k = \mathbb{C}$, and consider the curves given by the following parametric representations:

$$(R)\quad \begin{aligned} x &= t^{8} \\ y &= t^{10} + t^{13} \\ z &= t^{12} + t^{15} \end{aligned} \qquad\qquad (R^{*})\quad \begin{aligned} x^{*} &= t^{8} \\ y^{*} &= t^{10} + t^{13} \\ z^{*} &= t^{12} + 2\,t^{15} \ . \end{aligned}$$

They are trivially equisingular E.s.1., since the characteristic exponents of (R) and (R^{*}) agree.

On the other hand by looking at the Hamburger-Noether expansions, we have

$$r = 2\ ;\quad h=1\ ,\ n=8\ ,\ h_1=3\ ,\ n_1=2\ ,\ h_2=\infty\ ,\ n_2=1\ .$$

$$r^{*}=2\ ;\quad h^{*}=1\ ,\ n^{*}=8\ ,\ h_1^{*}=5\ ,\ n_1^{*}=2\ ,\ h_2^{*}=\infty\ ,\ n_2^{*}=1\ .$$

Thus, they are not equisingular E.s.2.

Conversely, with $k = \mathbb{C}$, the curves

$$(R')\quad \begin{aligned} x &= t^{8} \\ y &= t^{10} + t^{11} \\ z &= t^{15} \end{aligned} \qquad\qquad (R'^{*})\quad \begin{aligned} x^{*} &= t^{8} \\ y^{*} &= t^{10} + t^{13} \\ z^{*} &= t^{15} \ , \end{aligned}$$

are equisingular E.s.2. but they have not evidently equal characteristic exponents, so they are not equisingular E.s.1.

4. EQUISINGULARITY E.s.3. COINCIDENCE OF SEMIGROUPS OF VALUES.

As we saw in chapter IV, for plane algebroid curves, equiresolution is equivalent to coincidence of semigroups of values. For twisted curves this is not true, and therefore the coincidence of semigroups of values turns out to be considered as a new equisingularity criterion (E.s.3.). In this section, and according to [26] we shall prove that by using a graded ring associated with every curve, equisingularity E.s.3. may be characterized in a purely algebraic fashion.

Definition 5.4.1.- Let $\square$, $\square^*$ be algebroid curves over the algebraically closed field k. We shall say that $\square$ and $\square^*$ are equisingular E.s.3. when $S(\square) = S(\square^*)$, where S denotes semigroup of values.

It is evident that the minimal set of generators of the semigroup of values is a complete system of invariants for E.s.3.

Now, let us consider a curve $\square$ over k, and denote by $\overline{\square}$ the integral closure of $\square$ in its quotient field, by $\underline{v}$ the natural valuation, and by $\overline{m}$ the maximal ideal of $\overline{\square}$.

For any $i \geqslant 0$, we set

$$M^i = \overline{m}^i \cap \square = \{z \in \square \ / \ \underline{v}(z) \geqslant i\}.$$

We have the filtration

$$(1) \qquad \square = M^0 \supset M^1 \supset \ldots \supset M^i \supset \ldots$$

which defines the graded ring

$$gr_M(\square) = \bigoplus_{i=o}^{\infty} M^i/M^{i+1} \qquad .$$

Lemma 5.4.2.- Let $(\overline{\beta}_\nu)_{0 \leq \nu \leq g}$ be the minimal set of generators of $S(\square)$. Then we have an isomorphism of graded rings:

$$gr_M(\square) \cong k(t^{\overline{\beta}_0},\ldots,t^{\overline{\beta}_g}) .$$

Proof: Take a uniformizing $t \in \overline{\square}$. Since $\overline{\beta}_\nu \in S(\square)$ there exists $y_\nu \in \square$ such that

$$y_\nu = t^{\overline{\beta}_\nu} + y'_\nu \qquad , \qquad \underline{v}(y'_\nu) > \overline{\beta}_\nu .$$

If $gr_{(t)}\, k((t))$ is the graded ring given by the filtration of $\overline{\square}$:

$$\overline{\square} = \overline{m}^0 \supset \overline{m}^1 \supset \ldots \supset \overline{m}^i \supset \ldots$$

we have an injective homomorphism of graded rings,

$$H : gr_M(\square) \longrightarrow gr_{(t)} k((t)) = k(t),$$

since this filtration induces the filtration (1) over $\square$.

As $t^{\overline{\beta}_\nu} = In_M\, y_\nu$, $Im\, H \supset k(t^{\overline{\beta}_0},\ldots,t^{\overline{\beta}_g})$. Conversely, if $In_M\, z \in gr_M(\square)$, with $z \in \square$, then $\underline{v}(z) = a_0\, \overline{\beta}_0 + \ldots + a_g \overline{\beta}_g$, with $a_0,\ldots,a_g \geq 0$, so

$$H(In_M(z)) = c\, t^{a_0 \overline{\beta}_0} \ldots t^{a_g \overline{\beta}_g} \in k(t^{\overline{\beta}_0},\ldots,t^{\overline{\beta}_g}),$$

with $c \in k$.

Proposition 5.4.3.- Two curves $\square$ and $\square^*$ are equisingular E.s.3. if and only if there exists an isomorphism of graded rings

$$gr_M(\square) \cong gr_M(\square^*) .$$

Proof: If $S(\square) = S(\square^*)$ then by the above lemma,

$$gr_M(\square) \cong k\left(t^{\overline{\beta}_0}, \ldots, t^{\overline{\beta}_g}\right) \cong gr_M(\square^*) ,$$

where $(\overline{\beta}_\nu)_{0 \leq \nu \leq g}$ is the minimal set of generators of $S(\square)=S(\square^*)$. The converse is evident since $k\left(t^{\overline{\beta}_0}, \ldots, t^{\overline{\beta}_g}\right)$ determines the semigroup.

Remark 5.4.4.- There is no relation among the three equisingularity definitions. Recall that we have already seen in section 3, that neither E.s.1. implies E.s.2. nor E.s.2. implies E.s.1.

E.s.1. $\not\Rightarrow$ E.s.3. The curves over $\mathbb{C}$ with parametric representations (R) and (R^*) as in 5.3.4. are equisingular E.s.1., but they have not the same semigroup of values, since $23 = \underline{v}(xz-y^2)$ belongs to the first semigroup but it does not so to the second one.

E.s.2 $\not\Rightarrow$ E.s.3. It suffices to consider the curves over $\mathbb{C}$ given by

$$x = t^4 \qquad x' = t^4$$

$$y = t^5 \qquad y' = t^5$$

$$z = t^6 \qquad z' = t^7 .$$

E.s.3. $\not\Rightarrow$ E.s.1 (resp. E.s.2.). The curves over $\mathbb{C}$,

$$x = t^4 \qquad x' = t^4$$

$$y = t^6 + t^9 \qquad y' = t^6$$

$$z' = t^{15} ,$$

have the same semigroup $4\,\mathbb{Z}_+ + 6\,\mathbb{Z}_+ + 15\,\mathbb{Z}_+$, but they are neither equisingular E.s.1. nor E.s.2. as one may easily check.

REFERENCES

1. Abhyankar, S.S., "Inversion and invariance of characteristic pairs". Am. J. Math. 89 (1967), 363-372.

2. Ancochea, G., "Curvas algebraicas sobre cuerpos algebraicamente cerrados de característica cualquiera". Memoria de la Real Academia de Ciencias Exactas, Físicas y Naturales. Madrid (1947), 1-36.

3. Angermüller, G., "Die Wertehalbgruppe einer ebenen irreduziblen algebroiden Kurve". Math. Z. 153 (1977), 267-282.

4. Atiyah-Mac Donald,"Introduction to commutative algebra". Adison Wesley Public. Co. London (1969).

5. Bennet, B. M., "On the characteristic functions a local ring". Ann. Math. 91 (1970), 25-87.

6. Bourbaki, N., "Algèbre commutative". Hermann. Paris (1967).

7. Brauner, K., " Zur Geometrie der Funktionen Zweier Komplexen Veränderlichen". Abl. Math. Sem. Hamburg, 6 (1928), 1-54.

8. Burau, W., " Kennzeichnung der Schlauchknoten". Abh. Math. Sem. Hamburg, 9 (1928), 125-133.

9. Endler, O., "Valuation Theory". Springer-Verlag-Berlin. New York (1972).

10. Enriques, F.-Chisini,O., "Teoria geometrica delle equazioni e delle funzioni algebriche", vol.2, libro quarto. Bologna (1918).

11. Fulton, W., "Algebraic curves". Benjamin Inc. New York (1968).

12. García Rodeja, E., "Desarrollo de Hamburger-Noether". Actas de la 1ª Reunión de Matemáticos Españoles. Madrid (1961).

13. Hironaka, H., "Introduction to the theory of infinitely near singular points". Memorias de Matemática del Instituto Jorge Juan, CSIC, Madrid (1974).

14. Lê Dũng Trang, "Noeuds Algebriques". Ann. Inst. Fourier, 23. Grenoble (1972), 117-126.

15. Lejeune, M., "Sur l'equivalence de courbes algebroides planes. Coefficients de Newton". Thèse. Université Paris VII. Paris (1973).

16. Lipmann, J., "Absolute saturation of one-dimensional local rings". Am. J. Math. 97 (1975), 771-790.

17. Milnor, J., "Singular points of complex hypersurfaces", Ann. of Math. Stud. 61. Princeton, Univ. Press.(1968).

18. Moh, T.T., "On characteristic pairs of algebroid plane curves for characteristic p". Bull. Inst. Math. Sinica 1 (1973), 75-91.

19. Romo, C., "Resolución de singularidades de variedades algebroides sobre un cuerpo de característica arbitraria". Tesis. Madrid (1976).

20. Van der Waerden, B.L., "Einführung in die algebraische Geometrie". Verlag von J. Springer. Berlin (1939).

21. Vicente, J.L., "Singularidades de curvas algebroides alabeadas". Tesis. Madrid (1973).

22. Walker, R., "Algebraic curves". Dover Inc. New York (1962).

23. Wall, H. S., "Continued fractions". Van Nostrand. New York (1967).

24. Zariski, O., "Algebraic surfaces". Springer-Verlang-Berlin. New York (1972).

25. Zariski, O., "Contribution to the problem of equisingularity". C.I.M.E.Varenna. Septembre (1969).

26. Zariski, O., "Le probleme des modules pour les branches planes". Cours donné au Centre de Math. de l'Ecole Polytechnique. Paris (1973). Appendice de B. Teissier.

27. Zariski, O., "General Theory of saturation and saturated local rings".(I) Am. J. Math. 93 (1971), 573-648. (II) Am. J. Math. 93 (1971), 872-964. (III) Am. J. Math. 97 (1975), 415-502.

28. Zariski, O., "Studies in equisingularity". (I) Am. J. Math. 87 (1965), 507-535. (II) Am. J. Math. 87 (1965), 972-1006. (III) Am. J. Math. 90 (1968), 961-1023.

29. Zariski-Samuel, "Commutative Algebra". Van Nostrand. Princeton (1958).

INDEX

SYMBOLS

$\mathbb{Z}$, $\mathbb{R}$, $\mathbb{C}$, integer , real , complex numbers

$\mathbb{Z}_+$, nonnegative integers

$\mathbb{R}_+$, nonnegative reals

$\mathbb{Z}_+^{\mathbb{Z}_+}$, infinite sequences of nonnegative integers

$k[[X_1, \ldots, X_N]]$, power series ring

$k((t))$, power series field

Emb , embedding dimension

Spec , spectrum

Proj , projective scheme

Bl , blowing up

$\underline{\nu}$, order

gr , graded ring

$\sqrt{}$, the root of an ideal

e , multiplicity

(C , D) , intersection multiplicity

(m , n) , greatest common divisor

dim , dimension

l , length